Salters-Nuffield
Advanced Biology

www.heinemann.co.uk

✓ Free online support
✓ Useful weblinks
✓ 24 hour online ordering

01865 888058

Inspiring generations

Heinemann Educational Publishers
Halley Court, Jordan Hill, Oxford OX2 8EJ
Part of Harcourt Education

Heinemann is the registered trademark of
Harcourt Education Limited

© University of York Science Education Group 2005

First published 2005
Published as trial edition 2002

09 08 07
10 9 8 7 6 5

British Library Cataloguing in Publication Data is available
from the British Library on request.

ISBN 978 0 435 62857 4

Project editors: Angela Hall, Michael Reiss, Catherine Rowell and Anne Scott

Edited by Ruth Holmes

Index compiled by Laurence Errington

Designed and typeset by Bridge Creative Services Limited, Bicester, Oxon

Original illustrations © Harcourt Education Limited 2005

Illustrated by Roger Farrington and Hardlines Ltd

Printed and bound in China by China Translation & Printing Services Ltd.

Acknowledgements
The authors and publishers would like to thank the following for permission to use
copyright material: Figure 1.7 on page 6 © British Heart Foundation and reproduced
under license from BHF; Figure 2.4 on page 56 adapted from Fig 22.6b, page 444 in
N Campbell, L Mitchell and J Reece (1997), *Biology: Concepts and Connections*, 2nd
Edn © 1997 by The Benjamin/Cummings Publishing Company. Reprinted by permission
of Pearson Education, Inc; Extract on page 74 from *Life Story* by William Nicholson
(© William Nicholson 1987), reproduced by permission of PFD (www.pfd.co.uk) on
behalf of William Nicholson; Figure 3.31 on page 120 adapted from Fig 21.7, page 394
in N Campbell, J Reece and L Mitchell (1999), *Biology*, 5th Edn © 1999 by Benjamin/
Cummings, an imprint of Addison Wesley Longman, Inc. Reprinted by permission of
Pearson Education, Inc; Figure 4.5 on page 148 from M Roberts, G Monger and M Reiss
(2000), *Advanced Biology*, Nelson Thornes; Figure 4.45 on page 184 © Nature, Vol. 399,
30 June 1999, reproduced with permission; Table 4.4 A on page 198 reproduced with
permission of International Panel on Climate Change. All Crown Copyright material is
reproduced with the permission of the Controller of HMSO and the Queen's Printer
for Scotland.

Every effort has been made to contact copyright holders of material reproduced in this
book. Any omissions will be rectified in subsequent printings if notice is given to the
publishers.

For photograph acknowledgements, please see page 229

Contents

Contributors

Many people from schools, colleges, universities, industries and the professions have contributed to the Salters-Nuffield Advanced Biology project. They include the following.

Central team

Angela Hall (Project Officer), Nuffield Curriculum Centre
Michael Reiss (Director), Institute of Education, University of London
Catherine Rowell (Project Officer), University of York Science Education Group
Anne Scott (Project Officer), University of York Science Education Group

Sarah Codrington, Nuffield Curriculum Centre
Nancy Newton (Secretary), University of York Science Education Group

Advisory committee

Professor R McNeill Alexander FRS	University of Leeds
Dr Roger Barker	University of Cambridge
Dr Allan Baxter	GlaxoSmithKline
Professor Sir Tom Blundell FRS (Chair)	University of Cambridge
Professor Kay Davies CBE FRS	University of Oxford
Professor Sir John Krebs FRS	Food Standards Agency
Professor John Lawton FRS	Natural Environment Research Council
Professor Peter Lillford CBE	University of York
Dr Roger Lock	University of Birmingham
Professor Angela McFarlane	University of Bristol
Dr Alan Munro	University of Cambridge
Professor Lord Robert Winston	Imperial College of Science, Technology and Medicine

Authors

Topic 1
Glen Balmer	Watford Grammar School
Alan Clamp	Ealing Tutorial College
Ginny Hales	Cambridge Regional College
Gill Hickman	Ringwood School
Liz Hodgson	Greenhead College, Huddersfield

Topic 2
Jon Duveen	City & Islington College, London
Brian Ford	The Sixth Form College, Colchester
Steve Hall	King Edward VI School, Southampton
Pauline Lowrie	Sir John Deane's College, Northwich
Jamie Shackleton	Cambridge Regional College

Topic 3
Richard Fosbery	The Skinners School, Tunbridge Wells
Laurie Haynes	School of Biological Sciences, University of Bristol

Liz Jackson	King James's School, Knaresborough
Jenny Owens	Rye St Antony School, Oxford
Nick Owens	Oundle School, Peterborough

Topic 4
Susan Barker	Institute of Education, University of Warwick
Martin Bridgeman	Stratton Upper School, Biggleswade
Mark Colyer	Oxford College of Further Education
Barbara Geatrell	The Burgate School, Fordingbridge
Paul Heppleston	
Christine Knight	
Peter Lillford	Department of Biology, University of York
David Slingsby	Wakefield Girls High School
Mark Smith	Leeds Grammar School
Jane Wilson	Coombe Dean School, Plymouth
Mark Winterbottom	King Edward VI School, Bury St Edmunds

We would also like to thank the following for their advice and assistance:
John Holman	University of York
Andrew Hunt	Nuffield Curriculum Projects Centre
Jenny Lewis	University of Leeds
Martin Bridge	University College London
Glen Dawkins and John Newton	NERC Centre for Population Biology, Imperial College
J Phillip Grime	University of Sheffield
Kevin Southern	University of Liverpool, Royal Liverpool Children's Hospital, Alder Hey
Penny Rashbass	University of Sheffield

Sponsors

The Salters' Institute	Pfizer Limited
The Nuffield Foundation	Boots plc
The Wellcome Trust	ICI plc
Zeneca Agrochemicals	The Royal Society of Chemistry

About the SNAB course

Salters-Nuffield Advanced Biology (SNAB) is much more than just another A-level specification. It is a complete course with its own distinctive philosophy and it is supported by exciting teaching, learning and support materials. SNAB combines the key concepts underpinning biology today with the opportunity to gain the wider skills that biologists now need.

SNAB is the result of three years of piloting, funded by a number of organisations including the Salters' Institute and the Nuffield Foundation. Following the pilot, we have been able to incorporate student and teacher feedback into the course to make it even better.

A context-led approach

In SNAB you study biology through real-life contexts. For example, most A-level biology courses start with cell biology or biochemistry. We don't. We start with an account of Mark, a 15-year-old who had a stroke, and Peter, an adult who had a heart attack. You study the biological principles needed to understand what happened to Mark and Peter. You then go on from the details of their cases to look at the factors that make it more likely that any of us will suffer from a stroke or heart attack. All four AS topics use this context-led approach with a storyline or contemporary issue presented and biological principles introduced when required to aid understanding of the context.

Building knowledge through the course

In SNAB there is not, for example, a topic labelled 'biochemistry' containing everything you might need to know on carbohydrates, fats, nucleic acids and proteins. In SNAB you study the biochemistry of these large molecules bit by bit throughout the course when you need to know the relevant information for a particular topic. In this way information is presented in manageable chunks and builds on existing knowledge.

Activities as an integral part of the learning process

You really have to get actively involved in SNAB. Throughout this book you will find references to a wide variety of activities. Through these you will be learning practical and other techniques as well as developing a wide range of skills including, for instance, data analysis, critical evaluation of information, communication and collaborative working.

Within the electronic resources you will find animations on such things as protein synthesis, the cardiac cycle and cell division. These animations are designed to help you understand the more difficult bits of biology. The support sections will be useful if you need help with biochemistry, mathematics and statistics, ICT, study skills, the examination or coursework.

SNAB and ethical debate

With rapid developments in biological science, we are faced with an increasing number of challenging decisions. For example, the rapid advances in gene technology present ethical dilemmas: the horticulturalist and farmer may have to decide whether to grow a genetically modified (GM) variety of plant; we need to decide whether to buy GM products.

In SNAB you develop the ability to discuss and debate these types of biological issues. There is rarely a right or wrong answer; but you learn to justify your own decisions using ethical frameworks.

Exams and coursework

The exams in SNAB are much like any advanced biology exams but particularly reward your ability to reason biologically and to use what you have learned in new contexts. You won't just be repeating information you have learned off by heart. Most of the exam questions are structured ones, though as you go through the course you begin to do short essays, building up to longer ones. We believe that essay writing will be very useful for you if you go on to university or to any sort of job that requires you to be able to write reports. You can find out more about the coursework and examinations in the electronic exam and coursework support section and in the specification.

We think that SNAB is the most exciting and up-to-date advanced biology course around. Whatever your interests are – whether you want to do an AS course or go on to A2 and study a biological subject at university – we hope you enjoy the course.

Any questions?

If you have any questions or comments about the materials you can write to us at:

The Salters-Nuffield Advanced Biology Project
Science Education Group
University of York
Heslington
York
YO10 5DD

How to use this book

There are a number of features in the student books that we hope will help your learning. Some features will also help you to find your way around.

This AS book covers the four AS topics. These are shown in the contents list, which also shows you the page numbers for the main sections within each topic. There is an index at the back of the book to help you to find what you are looking for.

Main text

Key terms in the text are shown in **bold type**. These terms are defined in the interactive glossary on the software and can be found using the 'search glossary' feature.

There is an introduction at the start of each topic which provides a guide to the sort of things you will be studying in the topic.

There is an '**Overview**' box on the first spread of each topic, so you know which biological principles will be covered.

> **Overview** of the biological principles covered in this topic
>
> This topic will introduce the concept of risks to health. You will study the relative sizes of risks and how these are assessed. You will consider how we view different risks – our perception of risk. You will also look at how health risks may be affected by lifestyle choices.

Occasionally in the topics there are also '**Key biological principle**' boxes where a fundamental biological principle is highlighted.

> **Key biological principle:** Why have a heart and circulation?
>
> The heart and circulation have one primary purpose – to move substances around the body. In very small organisms, such as unicellular creatures, substances such as oxygen, carbon dioxide and digestive products are moved around the organism by diffusion. Diffusion is the relatively slow random movement of molecules.
>
> **Open circulatory systems**
> In insects and some other animal groups blood circulates in large open spaces. A simple heart pumps blood out of the arteries into cavities surrounding the animal's organs. Substances can diffuse between the blood and cells.

'**Did you know?**' boxes contain material that will not be examined, but we hope you will find interesting.

> **Did you know?** Why do we have such a sweet tooth?
>
> We have taste receptors on the tongue for five main tastes – sweet, sour, bitter, salty and umami (the taste associated with monosodium glutamate, MSG). It is likely that the sweet-taste receptors enable animals to identify food that is easily digestible.

Questions

You will find two types of question in this book.

In-text questions occur now and again in the text. They are intended to help you to think carefully about what you have read and to aid your understanding. You can self-check using the answers provided at the back of the book.

> blood is under low pressure in veins. Because of this the walls of veins are much thinner than those of arteries. Low pressure developed in the thorax (chest cavity) when breathing in also helps draw blood back into the heart.
>
> **Q1.5** List the features shown in Figure 1.10B that enable the artery to withstand high pressure and then recoil to maintain a steady flow of blood.
>
> Since the heart is a muscle it needs a constant supply of fresh blood. You might think that receiving a blood supply would never be a problem for the heart. However, the heart is unable to use any of the blood inside its pumping chambers directly. Instead, the heart muscle is supplied with blood through two vessels called the **coronary arteries**.
>
> **Checkpoint** ✓
>
> **1.2** Identify the key structures of an artery, a vein and a capillary, and in each case explain how the structure is related to the function of the vessel.

Boxes containing '**Checkpoint**' questions are found throughout the book. They give you summary-style tasks that build up some revision notes as you go through the student book.

Links to the software

Boxes that have an arrow icon link to features on the software. They are found in the margin near to the relevant piece of text.

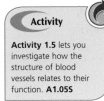

Activity

Activity 1.5 lets you investigate how the structure of blood vessels relates to their function. **A1.05S**

'**Activity**' boxes show you which activities are associated with particular sections of the book. Activity sheets and any related animations can be accessed from an activity homepage for each activity, found via 'topic resources' on the software. There may also be weblinks to useful websites. Activity sheets include such things as practicals, issues for debate and role plays. They can be printed out. Your teacher or lecturer will guide you on which activity to do and when.

A final activity for each topic enables you to **check your notes** using the topic summary provided within the activity. The summary shows you what you need to have learned from each topic for your unit exam.

Weblink

To check out the most recent death rate figures for coronary heart disease see the National Statistics Office website.

'**Weblink**' boxes give you useful websites to go and look at. They are provided on a dedicated 'weblinks' page on the software which is found under 'SNAB communications'. You can also access them through the Heinemann website at www.heinemann.co.uk/hotlinks.

Extension

Read **Extension 1.1** to find out how you may be able to save someone's life by carrying out cardiopulmonary resuscitation. **X1.01S**

'**Extension**' boxes refer you to extra information, associated with particular sections, which you can find on the software. The extension sheets can be printed out. The material in them will not be examined.

Support

To find out more about polar molecules and polar bonds have a look at the Biochemistry support on the website.

'**Support**' boxes are provided now and again, where it is particularly useful for you to go to the student support provision on the software, e.g. biochemistry support. You will also be guided to the support on the software from the activity homepages, or you can go there directly via 'student support'.

Review

Are you ready to tackle Topic 1 *Lifestyle, health and risk?*

Complete the GCSE review and GCSE review test before you start.

GCSE reviews and interactive GCSE review tests are provided for you to check that you can remember and understand enough from your GCSE studies to start a topic. You will find a '**Review**' box on the first spread of each topic.

Topic test

Now that you have finished Topic 1, complete the end-of-topic test before starting Topic 2.

At the end of each topic, as well as the **check your notes** activity for consolidation of the topic, there is a '**Topic test**' box for an interactive topic test. This test will usually be set by your teacher/lecturer, and will help you to find out how much you have learned from the topic.

Why a topic called Lifestyle, health and risk?

Congratulations on making it this far! Not everyone who started life's journey has been so lucky. In the UK only about 50% of conceptions lead to live births, and about 6 in every 1000 newborn babies do not survive their first year of life (Figure 1.1). After celebrating your first birthday there seem to be fewer dangers. Fewer than 2 in every 1000 children die between the ages of 1 and 14 years old. All in all, life *is* a risky business.

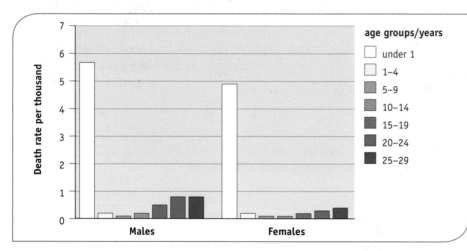

Figure 1.1 Death rates per 1000 population per year by age group and sex. Is life more risky for boys? *Source: England and Wales Office for National Statistics 2003.*

In everything we do there is some risk. Normally we only think something is risky if there is the obvious potential for a harmful outcome. Snowboarding, parachute jumping and taking ecstasy are thought of as risky activities, but even crossing the road, jogging or sitting in the sun have risks, and many people take actions to reduce them (Figures 1.2 and 1.3).

Risks to health are often not so apparent as the risks facing someone making a parachute jump. People often do not realise they are at risk from a lifestyle choice they make. They underestimate the effect such choices might have on their health.

What we eat and drink, and the activities we take part in, all affect our health and well-being. Every day we make choices that may have short- and long-term consequences of which we may be only vaguely aware. What are the health risks we are subjecting ourselves to? Will a cooked breakfast set us up for the day or will it put us on course for heart disease? Does the 10-minute walk to work really make a difference to our health?

Cardiovascular disease is the biggest killer in the UK, with one in three people dying from diseases of the circulatory system. Does everyone have the same risk? Can we assess and reduce the risk to our health? Do we need to? Is our perception of risk at odds with reality?

Figure 1.2 Some activities are less obviously risky than others, but may still have hidden dangers.

In this topic you will read about Mark and Peter, who have kindly agreed to share their experiences of cardiovascular disease. The topic will introduce the underlying biological concepts that will help you understand how cardiovascular diseases develop, and the ways of reducing the risk of developing these diseases.

Overview of the biological principles covered in this topic

This topic will introduce the concept of risks to health. You will study the relative sizes of risks and how these are assessed. You will consider how we view different risks – our perception of risk. You will also look at how health risks may be affected by lifestyle choices.

Building on your previous knowledge of the circulatory system you will study the heart and circulation and understand how these are affected by our choice of diet and activity.

You will look in some detail at the biochemistry of our food. This will give you a detailed understanding of some of the current thinking among doctors and other scientists about the choice of foods to reduce the risks to our health.

Review

Are you ready to tackle Topic 1 *Lifestyle, health and risk?*

Complete the GCSE review and GCSE review test before you start.

Figure 1.3 A UK male aged 15 to 24 is over three times more likely to have a fatal accident than a female of the same age.

Mark's story

On 28 July 1995 something momentous happened that changed my life...

▲ **Figure 1.4** Mark at 15.

I was sitting in my bedroom playing on my computer when I started to feel dizzy with a slight headache. Standing, I lost all balance and was feeling very poorly. I think I can remember trying to get downstairs and into the kitchen before fainting. People say that unconscious people can still hear. I don't know if it's true but I can remember my dad phoning for a doctor and that was it. It took 5 minutes from me being an average 15-year-old to being in a coma.

I was rushed to Redditch Alexandra Hospital where they did some reaction tests on me. They asked my parents questions about my lifestyle (did I smoke, take drugs, etc.?). Failing to respond to any stimulus, I was transferred in an ambulance to Coventry Walsgrave Neurological Ward. Following CT and MRI scans on my brain it was concluded that I had suffered a stroke. My parents signed the consent form for me to have an operation lasting many hours. I was given about a 30% chance of survival.

They stopped the bleed by clipping the blood vessels that had burst with metal clips, and removing the excess blood with a vacuum. I was then transferred to the intensive care unit to see if I would recover. Within a couple of days I was conscious and day by day I regained my sight, hearing and movement (although walking and speech were still distorted).

This is a true story. Mark had a stroke, one of the forms of cardiovascular disease. It is rare for someone as young as Mark to suffer a stroke. Why did it happen? Was he in a high-risk group?

▲ **Figure 1.5** The experience is not stopping Mark living life to the full.

Peter's story

I got the first indication of cardiovascular problems aged 23, when I was told that I had high blood pressure. I didn't really take much notice. My father had died at the age of 53 from a heart attack but as he was about four stone overweight, had a passion for fatty foods and smoked 60 full strength cigarettes a day, I didn't compare his condition to mine. I had a keen interest in sport, playing hockey and joining the athletics team at work. I was never overweight but I must admit that I probably drank too much at times and didn't bother too much about calories and cholesterol in food.

In 1981, I ran my first marathon at the age of 42 and subsequently did another five. All was going well I thought, until a routine medical showed my blood pressure reading to be 240 over 140. The doctor could not believe that I was still walking around, let alone running, and sent me straight to my GP. Since then I have always taken tablets for blood pressure and have also reviewed my diet.

I did continue running and completed the Great North Run at the age of 63. Thinking about doing the Great North Run again, I was running 8 miles a week and playing hockey. Then my eight-day holiday in Ireland became three days touring and twelve days in hospital.

At 2 o'clock in the morning I woke up with a terrific pain in my chest. I was sweating profusely and looking very pale. I had had a heart attack and within an hour I was in intensive care. At 5 am I had a second attack and the specialist inserted a temporary pacemaker to keep my heart rate up as it was dropping below 40.

After five days in intensive care I was transferred to the general ward for recuperation. I was told that it was possible that, had I not looked after myself, I might have had a heart attack much earlier in life.

On returning home I had an angiogram and was told that I needed a triple bypass operation. I have to say it was not pleasant, but I had decided that it was necessary and I would cope with anything that happened if it would get me back to a decent lifestyle. Well the operation, a quadruple bypass, was a success and after eight days I was back home.

This is a true story. Why did it happen to Peter, who seemed to be so active and healthy?

Figure 1.6 Peter's active lifestyle did not prevent his heart attack but probably helped him to make a full recovery.

Activity

To find out what happened to Mark and Peter read their full stories in **Activity 1.1**. **A1.01S**

1.1 What is cardiovascular disease?

Deaths from cardiovascular disease

Cardiovascular diseases (**CVDs**) are diseases of the heart and circulation. They are the main cause of death in the UK, accounting for over 250 000 deaths a year, and over 70 000 of these are premature deaths (Figure 1.7). More than one in three people in the UK die from cardiovascular diseases. The main forms of cardiovascular diseases are **coronary heart disease** (**CHD**) as experienced by Peter, and **stroke** as experienced by Mark.

About half of all deaths from cardiovascular diseases are from coronary heart disease and about a quarter are from stroke. Coronary heart disease is the most common cause of death in the UK. One in four men and one in five women die from the disease.

Weblink

To check out the most recent death rate figures for coronary heart disease see the National Statistics Office website.

Activity

Activity 1.2 demonstrates mass flow. **A1.02S**

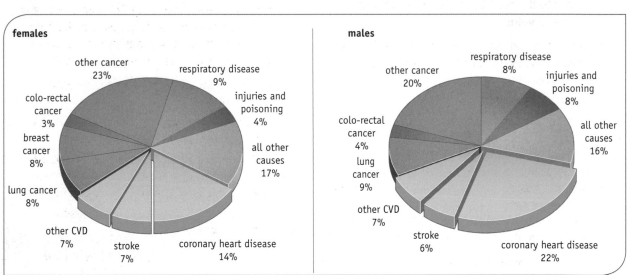

Figure 1.7 Premature deaths by cause in the UK in 2000 for females (left) and males (right). (Premature death is death under the age of 75 years.) One person dies of heart disease every 3 minutes. *Source: British Heart Foundation.*

Key biological principle: Why have a heart and circulation?

The heart and circulation have one primary purpose – to move substances around the body. In very small organisms, such as unicellular creatures, substances such as oxygen, carbon dioxide and digestive products are moved around the organism by diffusion. Diffusion is the relatively slow random movement of molecules.

Most complex multicellular animals, however, are too large for diffusion to move substances around their bodies quickly enough. These animals usually have blood to carry these vital substances around their bodies and a heart to pump it instead of relying on diffusion. In other words, they have a circulatory system.

Open circulatory systems
In insects and some other animal groups blood circulates in large open spaces. A simple heart pumps blood out of the arteries into cavities surrounding the animal's organs. Substances can diffuse between the blood and cells. When the heart muscle relaxes, blood is drawn from the cavity back into the heart, through small valved openings along its length. Some animals have more than one heart – the humble earthworm, for instance, has five.

Closed circulatory systems

Many animals, including all vertebrates, have a closed circulatory system in which the blood is enclosed within tubes. This generates higher blood pressure as the blood is forced along fairly narrow channels instead of flowing into large cavities. This means the blood travels faster and so the blood system is more efficient at delivering substances around the body:

- The blood leaves the heart under pressure and flows along **arteries** and then **arterioles** (small arteries) to **capillaries**.
- There are extremely large numbers of capillaries. These come into close contact with most of the cells in the body, where substances are exchanged between blood and cells.
- After passing along the capillaries, the blood returns to the heart by means of **venules** (small veins) and **veins**.

Valves ensure that blood flows only in one direction. Animals with closed circulatory systems are generally larger in size.

Single circulatory systems

Animals with a closed circulatory system have either single circulation or double circulation (Figure 1.8). Single circulation is found, for example, in fish:

- The heart pumps deoxygenated blood to the gills.
- Here gaseous exchange takes place: there is a net diffusion of carbon dioxide from the blood into the water that surrounds the gills, and a net diffusion of oxygen from this water into the blood.

- The blood leaving the gills then flows round the rest of the body before eventually returning to the heart.

Note that the blood flows through the heart once for each complete circuit of the body.

Double circulatory systems

Birds and mammals, though, have double circulation:

- The right ventricle of the heart pumps deoxygenated blood to the lungs to receive oxygen.
- The oxygenated blood then returns to the heart to be pumped a second time (by the left ventricle) out to the rest of the body.

This means that the blood flows through the heart twice for each complete circuit of the body. The heart gives the blood returning from the lungs an extra 'boost'. This reduces the time it takes for the blood to circulate round the whole body. This allows birds and mammals to have a higher metabolic rate, because oxygen and food substances required for metabolic processes can be delivered more rapidly to cells.

Q1.1 Why do only small animals have an open circulatory system?

Q1.2 What are the advantages of having a double circulatory system?

Q1.3 Fish have two-chamber hearts and mammals have four-chamber hearts. Sketch what the three-chamber heart of an amphibian might look like.

Q1.4 What might be the major disadvantage of this three-chamber system?

Checkpoint

1.1 Make a bullet point summary which explains why many animals have a heart and circulation.

Activity

Activities 1.3 and **1.4** let you look in detail at the structure of a mammalian heart using either a dissection or a simulation. **A1.03S** (actual dissection) **A1.04S** (simulated dissection)

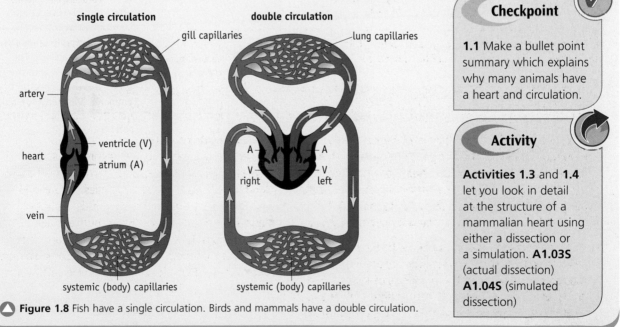

Figure 1.8 Fish have a single circulation. Birds and mammals have a double circulation.

How does the circulation work?

The heart and blood vessels

Study Figure 1.9 and locate the arteries carrying blood away from the heart and the veins returning blood to the heart.

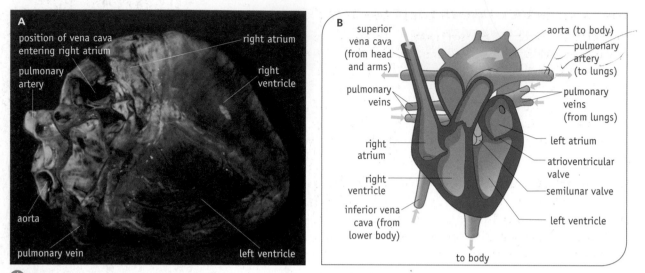

▲ **Figure 1.9 A** A normal human heart (dorsal or back view). **B** Diagrammatic cross-section of the human heart (ventral or front view).

Arteries and veins can easily be distinguished, as shown in Figure 1.10. The walls of both vessels contain **collagen**, a tough fibrous protein, which makes them tough and durable. They also contain elastic fibres and smooth muscle cells, which allows them to stretch and recoil. The key differences between the arteries and veins are listed below.

Arteries:
- narrow lumen
- thicker walls
- more collagen, elastic fibres and smooth muscle
- no valves

Veins:
- wide lumen
- thinner walls
- less collagen, elastic fibres and smooth muscle
- valves

▼ **Figure 1.10**
A Photomicrograph of an artery (left) and vein (right) surrounded by connective tissue. **B** Diagram of an artery, a vein and a capillary. The endothelium that lines the blood vessels is made up of epithelial cells (see page 53).

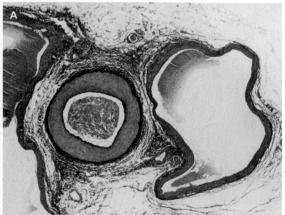

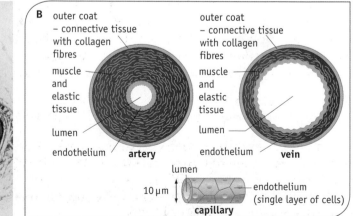

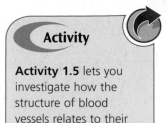

The capillaries that join the small arteries (arterioles) and small veins (venules) are very narrow, about $10\,\mu m$ in diameter, with thin walls only one cell thick.

These features can be directly related to the functions of the blood vessels, as described below.

How does blood move through the vessels?

Every time the heart contracts (**systole**), blood is forced into arteries and their elastic walls stretch to accommodate the blood. During **diastole** (relaxation of the heart), the elasticity of the artery walls causes them to contract behind the blood, pushing the blood forward. The blood moves along the length of the artery as each section in series is stretched and recoils in this way. This maintains a continual, pulsing flow of blood through the arteries. This pulse of blood moving through the artery can be felt anywhere an artery passes over a bone close to the skin.

By the time the blood reaches the smaller arteries and capillaries there is a steady flow of blood. In the capillaries this allows exchange between blood and the surrounding cells through the one-cell thick capillary walls. The network of capillaries that lies close to every tissue cell ensures that there is rapid diffusion between the blood and surrounding cells.

Activity

Activity 1.5 lets you investigate how the structure of blood vessels relates to their function. **A1.05S**

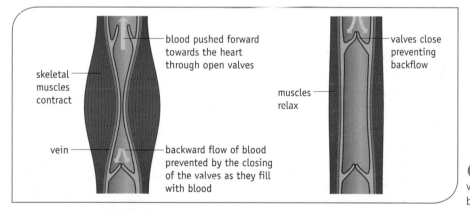

skeletal muscles contract

blood pushed forward towards the heart through open valves

vein

backward flow of blood prevented by the closing of the valves as they fill with blood

muscles relax

valves close preventing backflow

Figure 1.11 Valves in the veins prevent the backflow of blood.

The heart has a less direct effect on the flow of blood through the veins. In the veins blood flow is assisted by the contraction of skeletal muscles during movement of limbs and breathing. Backflow is prevented by valves within the veins (Figure 1.11). The steady flow without pulses of blood means that the blood is under low pressure in veins. Because of this the walls of veins are much thinner than those of arteries. Low pressure developed in the thorax (chest cavity) when breathing in also helps draw blood back into the heart.

Q1.5 List the features shown in Figure 1.10B that enable the artery to withstand high pressure and then recoil to maintain a steady flow of blood.

Since the heart is a muscle it needs a constant supply of fresh blood. You might think that receiving a blood supply would never be a problem for the heart. However, the heart is unable to use any of the blood inside its pumping chambers directly. Instead, the heart muscle is supplied with blood through two vessels called the **coronary arteries**.

Checkpoint

1.2 Identify the key structures of an artery, a vein and a capillary, and in each case explain how the structure is related to the function of the vessel.

9

How does the heart work?

Give a tennis ball a good, hard squeeze. You're using about the same amount of force that your heart uses in a single contraction to pump blood out to the body. Even when you are at rest, the muscles of your heart work hard – weight for weight, harder than the leg muscles of a person running.

The chambers of the heart alternately contract (systole) and relax (diastole) in a rhythmic cycle. One complete sequence of filling and pumping blood is called a **cardiac cycle**, or heartbeat. During systole, cardiac muscle contracts and the heart pumps blood out through the aorta and pulmonary arteries. During diastole, cardiac muscle relaxes and the heart fills with blood.

The cardiac cycle can be simplified into three phases: atrial systole, ventricular systole and diastole. The events that occur during each of the stages are shown in Figure 1.12.

Phase 1: Atrial systole

Blood returns to the heart due to the action of skeletal and gaseous exchange (breathing) muscles as you move and breathe. Blood under low pressure flows into the **left** and **right atria**. As the atria fill, the pressure of blood against the **atrioventricular valves** pushes them open and blood begins to leak into the **ventricles**. The atria walls contract, forcing more blood into the ventricles. This is known as atrial systole.

Phase 2: Ventricular systole

Atrial systole is immediately followed by ventricular systole. The ventricles contract from the base of the heart upwards, increasing the pressure in the ventricles. This pushes blood up and out through the arteries. The pressure of blood against the atrioventricular valves closes them and prevents blood flowing backwards into the atria.

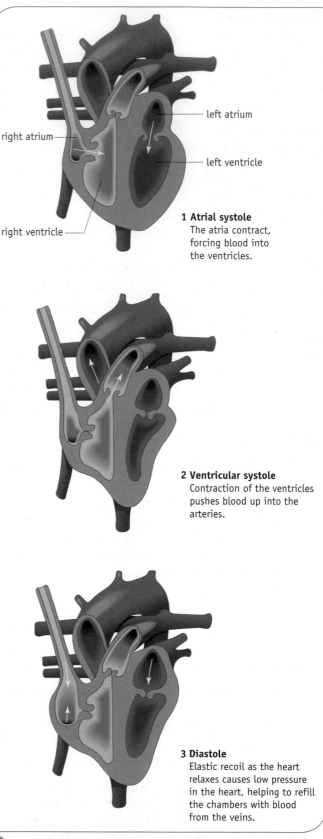

right atrium

left atrium

left ventricle

right ventricle

1 Atrial systole
The atria contract, forcing blood into the ventricles.

2 Ventricular systole
Contraction of the ventricles pushes blood up into the arteries.

3 Diastole
Elastic recoil as the heart relaxes causes low pressure in the heart, helping to refill the chambers with blood from the veins.

Figure 1.12 The three stages of the cardiac cycle. At each stage blood moves from higher to lower pressure.

Phase 3: Diastole

The atria and ventricles then relax during diastole. Elastic recoil of the relaxing heart walls lowers pressure in the atria and ventricles. Blood under higher pressure in the arteries is drawn back towards the ventricles, closing the **semilunar valves** and preventing further backflow. Low pressure in the atria helps draw blood into the heart from the veins.

Closing of the atrioventricular valves and then the semilunar valves creates the characteristic sounds of the heart.

Q1.6 When the heart relaxes in diastole, you might expect blood to move from the arteries back into the ventricles due to the action of gravity and the elastic recoil of the heart. How is this prevented?

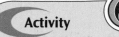

Activity

Activity 1.6 lets you test your knowledge of the cardiac cycle.
A1.06S

Checkpoint

1.3 Make a flowchart which summarises the events in the cardiac cycle.

What is atherosclerosis?

Atherosclerosis is the disease process that leads to coronary heart disease and strokes. In atherosclerosis fatty deposits can either block an artery directly, or increase its chance of being blocked by a blood clot (**thrombosis**). The blood supply can be blocked completely. If this happens for long, the affected cells are permanently damaged. In the arteries supplying the heart this results in a heart attack (**myocardial infarction**); in the arteries supplying the brain it results in a stroke. The supply of blood to the brain is restricted or blocked, causing damage or death to cells in the brain. Narrowing of arteries to the legs can result in tissue death and gangrene (decay).

What happens in atherosclerosis?

Atherosclerosis can be triggered by a number of factors. Whatever the trigger, this is the course of events that follows:

1 The delicate layer of cells that lines the inside of an artery (Figure 1.13A), separating the blood that flows along the artery from the muscular wall, becomes damaged for some reason. For instance, this damage can result from high blood pressure, which puts an extra strain on the layer of cells, or it might result from some of the toxins from cigarette smoke in the bloodstream.

2 Once the inner lining of the artery is breached, there is an inflammatory response. Large white blood cells leave the blood vessel and move into the artery wall. These cells accumulate chemicals from the blood, particularly **cholesterol**. A deposit builds up, called an **atheroma**.

3 Calcium salts and fibrous tissue also build up at the site, resulting in a hard swelling called a **plaque** on the inner wall of the artery. The build-up of fibrous tissue means that the artery wall loses some of its elasticity: in other words, it hardens. The ancient Greek word for 'hardening' is 'sclerosis', giving the word 'atherosclerosis'.

4 Plaques cause the artery to become narrower (Figure 1.13B). This makes it more difficult for the heart to pump blood around the body and can lead to a rise in blood pressure. Now there is a dangerous **positive feedback** building up. Plaques lead to raised blood pressure and raised blood pressure makes it more likely that further plaques will form.

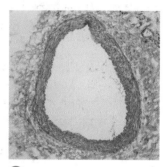

▲ **Figure 1.13A**
Photomicrograph of a normal, healthy coronary artery showing no thickening of the arterial wall. The lumen is large. Magnification ×215.

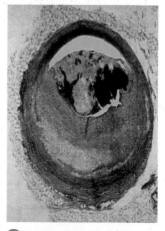

▲ **Figure 1.13B**
Photomicrograph of a diseased coronary artery showing narrowing of the lumen due to build up of atherosclerotic plaque. Magnification ×230.

The person is probably unaware of any problem at this stage, but if the arteries become very narrow or completely blocked then they cannot supply enough blood to bring oxygen and nutrients to the tissue. The tissue can no longer function normally and symptoms will soon start to show.

Why does the blood clot in arteries?

When blood vessel walls are damaged or blood flows very slowly, a blood clot is much more likely to form (Figure 1.13C). **Platelets** are red blood cell fragments without a nucleus which are present in blood. When these come into contact with the vessel wall they change from flattened discs to spheres with long thin projections. Their cell surfaces change, causing them to stick to the exposed collagen in the wall and to each other to form a platelet plug. They release **ADP** and other substances that activate more platelets.

Any damage of the vessel wall brings blood into direct contact with collagen within the wall. This triggers a complex series of chemical changes in the blood. A cascade of changes results in the soluble **plasma protein** called **prothrombin** being converted into **thrombin**. Thrombin is an enzyme that catalyses the conversion of another soluble plasma protein, **fibrinogen**, into long insoluble strands of the protein **fibrin**. These fibrin strands form a tangled mesh that traps blood cells to form a clot (Figures 1.14 and 1.15).

Why do only arteries get atherosclerosis?

The fast-flowing blood in arteries is under high pressure so there is a significant chance of damage to the walls. The low pressure in the veins means that there is little risk of damage to the walls.

The symptoms of cardiovascular disease

Stroke

The effects of a stroke will vary depending on the type of stroke, where in the brain the problem has occurred and the extent of the damage. The more extensive the damage the more severe the stroke and the lower the chance of full recovery. The symptoms normally appear very suddenly and include:

- numbness
- dizziness
- confusion
- slurred speech
- blurred or loss of vision, often only in one eye.

Visible signs often include paralysis on one side of the body with a drooping arm, leg or eyelid, or a dribbling mouth. The *right* side of the brain controls the *left* side of the body, and vice versa; therefore the paralysis occurs on the opposite side of the body to where the stroke occurred.

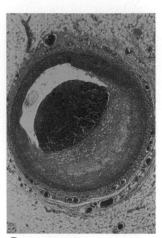

▲ **Figure 1.13C**
Photomicrograph of a diseased coronary artery showing narrowing and a blood clot. Magnification ×245.

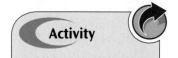

Activity

Activity 1.7 lets you summarise the steps in development of atherosclerosis and clot formation. **A1.07S**

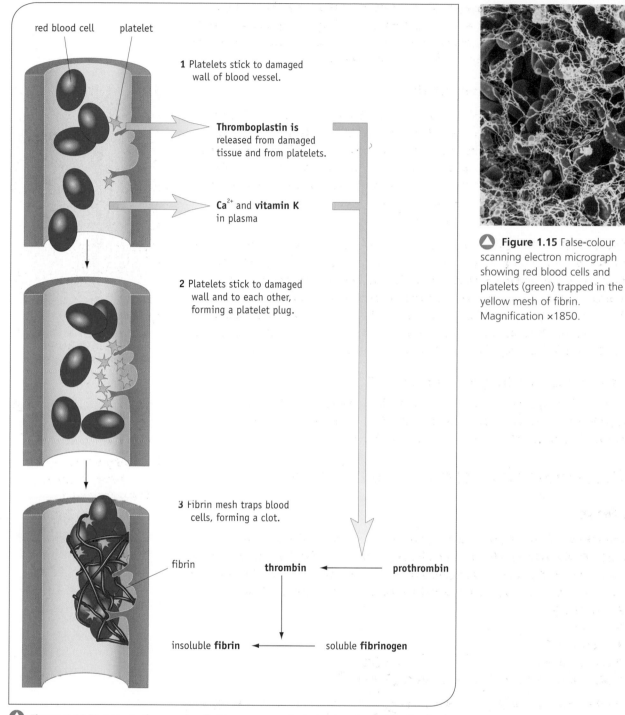

red blood cell   platelet

1 Platelets stick to damaged wall of blood vessel.

Thromboplastin is released from damaged tissue and from platelets.

Ca^{2+} and **vitamin K** in plasma

2 Platelets stick to damaged wall and to each other, forming a platelet plug.

3 Fibrin mesh traps blood cells, forming a clot.

fibrin

thrombin ◄── **prothrombin**

insoluble **fibrin** ◄── soluble **fibrinogen**

Figure 1.14 Damage to the vessel walls triggers a complicated series of reactions that leads to clotting.

Figure 1.15 False-colour scanning electron micrograph showing red blood cells and platelets (green) trapped in the yellow mesh of fibrin. Magnification ×1850.

If the supply of blood to the brain is only briefly interrupted then a mini-stroke may occur, called a transient ischaemic attack. A transient ischaemic attack has all the symptoms of a full stroke but the effects last for only a short period, and full recovery can happen quite quickly. However, a transient ischaemic attack is a warning of problems with blood supply to the brain that could result in a full stroke in the future.

Coronary heart disease

Angina

Narrowing of the coronary arteries limits the amount oxygen-rich blood reaching the heart muscle. The result may be a chest pain called **angina**. Angina is usually experienced during exertion. Because the heart muscle lacks oxygen it is forced to respire **anaerobically**. This produces **lactic acid** and the accumulation of lactic acid causes the pain of angina. Intense pain, an ache or a feeling of constriction and discomfort is felt in the chest or in the left arm and shoulder.

Shortness of breath and angina are often the first signs of coronary heart disease. Other symptoms are unfortunately very similar to those of severe indigestion and include a feeling of heaviness, tightness, pain, burning and pressure – usually behind the breastbone, but sometimes in the jaw, arm or neck.

Myocardial infarction (heart attack)

If a fatty plaque in the coronary arteries ruptures, cholesterol is released which leads to rapid clot formation. The blood supply to the heart may be blocked completely (Figure 1.16). The heart muscle supplied by these arteries does not receive any blood, so it is said to be **ischaemic** (without blood). If the affected muscle cells are starved of oxygen for long they will be permanently damaged. This is what we call a heart attack or myocardial infarction. If the zone of dead cells occupies only a small area of tissue the heart attack is less likely to prove fatal.

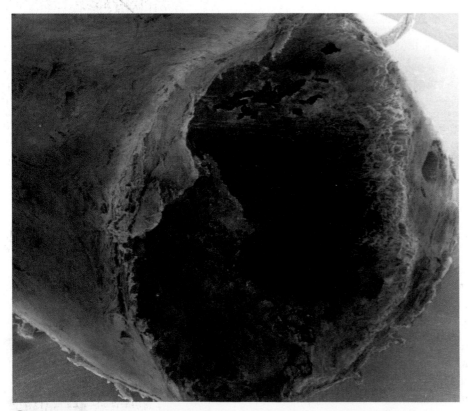

Figure 1.16 A small clot can block one of the coronary arteries. Magnification ×50.

Extension

Read **Extension 1.1** to find out how you may be able to save someone's life by carrying out cardiopulmonary resuscitation. **X1.01S**

Checkpoint

1.4 Draw up a table that lists the main symptoms of cardiovascular disease. Include both coronary heart disease and stroke.

Sometimes coronary heart disease causes the heart to beat irregularly. This is known as **arrhythmia** and can itself lead to heart failure. It can be important in the diagnosis of coronary heart disease and you can read more details on page 19.

Did you know? Aneurysms

If part of an artery has narrowed and become less flexible, blood can build up behind it. The artery bulges as it fills with blood and an **aneurysm** forms. An atherosclerotic aneurysm of the aorta is shown in Figure 1.17.

What will eventually happen as the bulge enlarges and the walls of the aorta are stretched thin? Aortic aneurysms are likely to rupture when they reach about 6–7 cm in diameter. The resulting blood loss and shock can be fatal. Fortunately earlier signs of pain may prompt a visit to the doctor. The bulge can often be felt in a physical examination or seen with ultrasound examination and it may be possible to surgically replace the damaged artery with an artificial artery.

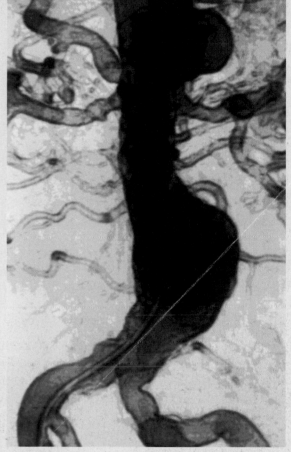

Figure 1.17 An aneurysm in the aorta below the kidneys. If an aneurism ruptures it can be fatal. Magnification ×0.8.

How is CVD diagnosed?

If a patient is suspected of having cardiovascular disease (CVD), the doctor will examine them, review their medical history and ask for tests to be carried out. The most common test to check for problems with the heart is an **electrocardiogram** (**ECG**). An electrocardiogram is a graphic record of the electrical activity of the heart as it contracts and rests. The test is easy to carry out and any patient with suspected cardiac problems will probably have one.

An electrocardiogram is also used on a patient with a suspected stroke to check for any heart condition that would increase the likelihood of stroke. If a stroke is suspected, as in Mark's case, a computerised axial tomography (CAT) scan and a magnetic resonance imaging (MRI) scan might be performed. These techniques are covered in detail in Topic 8.

Extension

There are several other tests to diagnose cardiovascular disease that can be requested by doctors and you can read more details of these tests in **Extension 1.2. X1.02S**

CT or CAT scan

A computerised axial tomography scan, normally referred to as a CT or CAT scan, uses X-rays. A series of X-rays is taken as the scanner rotates around the patient and the computer generates a 3D image from these.

MRI

In a magnetic resonance imaging scan the patient is placed within a magnetic field and radio waves are used to create images of the brain or other parts of the body. The energy in the radio signals is absorbed and then re-emitted by atoms within the body. The rate of emission is detected and is dependent on tissue type. The computer generates a series of black-and-white pictures representing different cross-sections of the brain or other body part. MRI images can therefore quickly reveal evidence of a stroke: areas of damaged brain tissue, aneurysms, internal bleeding or blockages.

MRI is also used for breast scans and the visualisation of injuries.

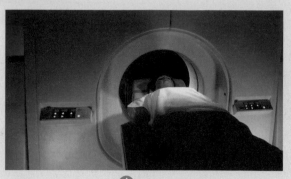

△ **Figure 1.18** Mark was diagnosed using both CAT and MRI scans. This photograph shows a patient about to enter a CAT scanner.

How does the heart beat?

The heart will beat without any input from the nervous system and it continues to beat, even outside the body, as long as its cells are alive. How does it do this?

Contraction of heart muscle is initiated by small changes in the electrical charge of heart cells. When these cells have a slight positive charge on the outside, they are said to be polarised. When this charge is reversed, they are depolarised. A change in polarity spreads like a wave from cell to cell and causes the cells to contract.

Depolarisation starts at the **sinoatrial node** (**SAN**). This is a small area of specialised muscle fibres located in the wall of the right atrium, beneath the opening to the superior vena cava (Figure 1.19). The sinoatrial node is also known as the **pacemaker**. The SAN generates an electrical **impulse** and this spreads across the right and left atria as a wave of depolarisation, causing them to contract at the same time. The impulse also travels to some specialised bundles of cells which take it to the **atrioventricular node** (**AVN**). From here the impulse is conducted to the ventricles after a delay of about 0.13 s.

Why is it important that the impulse is delayed? The delay ensures that the atria have finished contracting and that the ventricles have filled with blood before they contract.

After this delay, the signal reaches the **Purkyne fibres** (sometimes called Purkinje fibres), large specialised muscle fibres that conduct impulses rapidly to the apex (tip) of the ventricles. There are right and left bundles of fibres and they are collectively called the **bundle of His**.

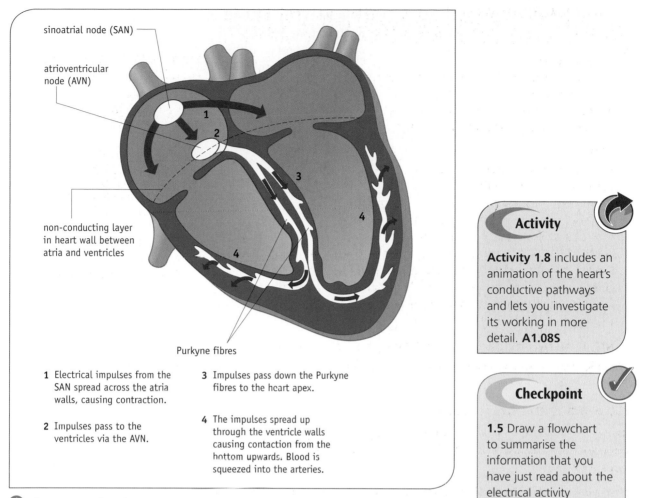

sinoatrial node (SAN)

atrioventricular node (AVN)

non-conducting layer in heart wall between atria and ventricles

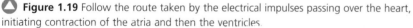

Purkyne fibres

1 Electrical impulses from the SAN spread across the atria walls, causing contraction.

2 Impulses pass to the ventricles via the AVN.

3 Impulses pass down the Purkyne fibres to the heart apex.

4 The impulses spread up through the ventricle walls causing contaction from the bottom upwards. Blood is squeezed into the arteries.

Figure 1.19 Follow the route taken by the electrical impulses passing over the heart, initiating contraction of the atria and then the ventricles.

Activity

Activity 1.8 includes an animation of the heart's conductive pathways and lets you investigate its working in more detail. **A1.08S**

Checkpoint

1.5 Draw a flowchart to summarise the information that you have just read about the electrical activity of the heart.

The Purkyne fibres continue around each ventricle and divide into smaller branches that penetrate the ventricular muscle. These branches carry the impulse to the inner cells of the ventricles and from here it spreads through each entire ventricle.

The first ventricular cells to be depolarised are at the apex of the heart, so that contraction begins at this point and travels *upwards* towards the atria. This produces a wave of contraction moving up the ventricles, pushing the blood into the aorta and the pulmonary artery.

How is an ECG carried out?

In an ECG, leads are attached to the person's chest and limbs to record the electrical currents produced during the cardiac cycle (see Figure 1.20). When there is a change in **polarisation** a small electrical current can be detected at the skin's surface. An ECG measures electrical current at the skin surface.

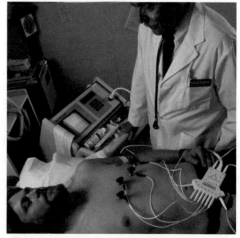

Figure 1.20 Twelve electrodes are used to give 12 views of the heart. By using different combinations of electrodes, the ECG can detect electrical currents as they spread in different directions across different regions of the heart.

An ECG is usually performed while the patient is at rest, lying down and relaxed, but it may be used in a stress test (a treadmill test). This test records the heartbeat during exercise and is used to detect heart problems that emerge only when the heart is working hard. The stress test involves doing an ECG before, during and after a period of exercise on a treadmill. Breathing rate and blood pressure may also be measured and recorded.

What does the ECG trace show us?

Look at Figure 1.21 and note which wave on the ECG represents each stage in the electrical activity of the heart.

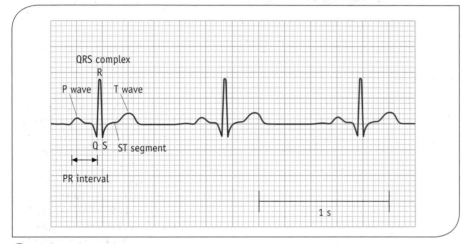

▲ **Figure 1.21** A normal ECG trace. The vertical axis shows electrical activity; the horizontal axis shows time.

- **P wave** – depolarisation of the atria that leads to atrial contraction (atrial systole)
- **PR interval** – the time taken for impulses to be conducted from the SAN across the atria to the ventricles, through the AVN
- **QRS complex** – the wave of depolarisation that results in contraction of the ventricles (ventricular systole)
- **T wave** – repolarisation (recovery) of the ventricles during the heart's relaxation phase (diastole)

The ECG does not show atrial repolarisation because the signals generated are small and are hidden by the QRS complex.

Using the ECG in diagnosis

The ECG can be used to measure heart rate. Each larger square represents 0.2 s; five squares represent 1 s (Figure 1.21) and 300 squares pass through the ECG machine in 1 minute. You can work out the time for one complete cardiac cycle by multiplying the number of squares between QRS complexes by 0.2 (the length in seconds of a square) and then divide this into 60. Alternatively you can divide 300 by the number of squares per beat. The two methods give the same answer.

Q1.7 Work out the time for one cardiac cycle, and the heart rate for the patient whose ECG trace is shown in Figure 1.21.

A heart rate of less than 60 beats per minute (60 bpm) is known as bradycardia. It is common in fit athletes at rest but can also be a symptom of heart problems. Possible causes include hypothermia, ischaemic heart disease or use of medicines or drugs.

Tachycardia, a heart rate greater than 100 bpm, is normally the result of anxiety, fear, fever or exercise. It can also be a symptom of coronary heart disease, heart failure, use of medicines or drugs, fluid loss or anaemia.

During a period of ischaemia, the normal electrical activity and rhythm of the heart are disrupted, and arrhythmias (irregular beatings caused by electrical disturbances) can affect a larger area of heart muscle than that affected by the initial ischaemia (Figure 1.22A).

An ECG trace can provide information about abnormal heartbeats, areas of damage and inadequate blood flow. Look at Figure 1.22 and then try answering the questions that follow.

> **Activity**
>
> Can you work out the patients' heart problems from the ECGs in **Activity 1.9**? **A1.09S**

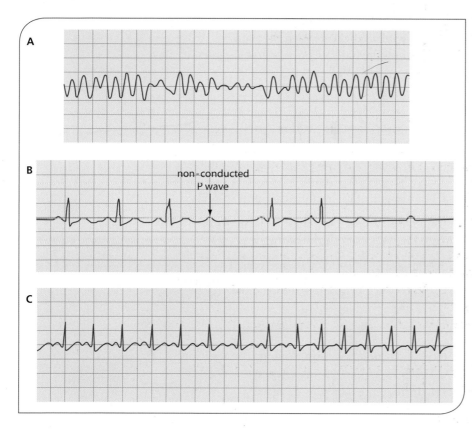

◀ **Figure 1.22 A** In ventricular fibrillation, irregular stimulation of the ventricles makes them contract in a weak and uncoordinated manner. This leads to a fall in blood pressure and results in sudden death unless treated immediately. **B** Notice that the P wave is not followed by a QRS complex, showing there is a break in the conduction system of the heart. **C** This ECG trace shows abnormal rhythms that could lead to cardiac arrest.

Q1.8 How does the trace in **a** Figure 1.22A and **b** Figure 1.22C differ from that of a normal ECG?

Q1.9 In Figure 1.22B where do you think that the break in the conduction system occurs?

Q1.10 The ECG in Figure 1.22C is from a young woman who had collapsed at a club. Can you suggest what might have happened to produce these rapid rhythms?

1.2 Who is at risk of cardiovascular disease?

Probability and risk

What do we mean by risk?

Risk is defined as 'the probability of occurrence of some unwanted event or outcome'. It is usually in the context of hazards, that is, anything that can potentially cause harm, such as the chance of contracting lung cancer if you smoke. Probability has a precise mathematical meaning and can be calculated to give a numerical value for the size of the risk. Do not panic – the maths is simple!

Taking a risk is a bit like throwing a die (singular of 'dice'). You can calculate the chance that you will have an accident or succumb to a disease (or throw a six). You will not *necessarily* suffer the accident or illness, but by looking at past circumstances of people who have taken the same risk, you can estimate the chance that you will suffer the same fate to a reasonable degree of accuracy.

Working out probabilities

There are six faces on a standard die. Only one face has six dots, so the chance of throwing a six is 1 in 6 (provided the die is not loaded). Scientists tend to express '1 in 6' as a decimal: 0.166 666 recurring (about 0.17). In other words, each time you throw a standard die, you have about a 0.17 or 17% chance of throwing a one, about a 17% chance of throwing a two, and so on.

When measuring risk you must always quote a time period for the risk. Here you have a 17% chance of throwing a one with each throw of the dice.

In a Year 5 class of 30 pupils, six children caught head lice in one year. The risk of catching head lice in this class was, therefore, 6 in 30 or 1 in 5, giving a probability of 0.2 or 20% in a year.

Estimating risks to health

In 1998, 19 523 people in the UK died due to injuries or poisoning. The total UK population at the time was 59 236 500 so we can calculate the average risk in a year of someone in the UK dying from injuries or poisoning as:

19 523 in 59 236 500

or 1 in $\dfrac{59\,236\,500}{19\,523}$

= 1 in 3034

$(= \dfrac{1}{3034})$

= 0.000 33 or 0.033%

Another way of working this out is as:

$$\frac{19\,523}{59\,236\,500}$$

However, when calculating a probability in relation to health, most people would find, for example, 1 in 3034 more meaningful than 0.000 33 or 0.033%.

Assuming the proportion of people that die from accidents or poisoning remains much the same each year, this calculation gives an estimate of the risk for any year.

If we calculated the risk of any one of us dying from lung cancer we would find a probability of 1 in about 1700. However, because lung cancer is much more likely if you smoke, the risk for smokers is far greater. When looking at calculated risk values you need to think about exposure to the hazard.

Q1.11   Look at the causes of death listed below and put them in order, from the most likely to the least likely. You could also have a go at estimating the probability of someone in the UK dying from each cause during a year.
- accidental poisoning
- heart disease
- injury purposely inflicted by another person
- lightning
- lung cancer
- railway accidents
- road accidents

Did you get it right?

People frequently get it wrong, underestimating or overestimating risk. We can say that there is a about a 1 in 1700 risk of each of us dying from lung cancer in any one year, a 1 in 100 000 risk of our being murdered in the next 12 months, and a 1 in 10 million risk of our being hit by lightning in a year. However, recent work on risk has concentrated not so much on numbers such as these but on the perception of risk.

Perception of risk

The significance of the perception of risk can be illustrated by a decision in September 2001 made by the American Red Cross, which provides about half of the USA's blood supplies. They decided to ban all blood donations from anyone who has spent six months or more in any European country since 1980. Their reason was the risk of transmitting variant Creutzfeldt–Jakob disease (vCJD), the human form of bovine spongiform encephalopathy (BSE), through blood transfusion. Experts agreed that there was a *chance* of this happening. Yet there wasn't a single known case of its *actually* having happened. Indeed, as the USA is short of blood for blood transfusions it is possible that more people may have died as a result of this 'safety precaution' than would have been the case without it.

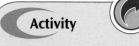

Activity

Activity 1.10 asks you to estimate risks for a range of diseases using the National Office for Statistics data on causes of death in the UK.
A1.10S

So why did America ban European blood donations? The likely reason was public perceptions of the risk of contracting vCJD. People will *overestimate* the risk of something happening if the risk is:

- involuntary (not under their control)
- not natural
- unfamiliar
- dreaded
- unfair
- very small.

If you look at this list you should be able to see why people may greatly overestimate some risks (such as the chances of contracting vCJD from blood transfusions) while underestimating others (such as the dangers of driving slightly faster than the speed limit or playing on a frozen lake). A useful distinction is sometimes made between risk and uncertainty. When we lack the data to estimate a risk precisely, we are *uncertain* about the risk. For example, we are uncertain about the environmental consequences of many chemicals.

Nowadays many risk experts argue that perceptions of risk are what really drive people's behaviour. Consider what happened when it became compulsory in the UK to use seat belts for children in the rear seats of cars (Figure 1.23). The number of children killed and injured *increased*. How could this be? John Adams, an academic at University College London, argues that this is because the parents driving felt safer once their children were wearing seat belts and so drove slightly less carefully. Unfortunately, this change in their driving behaviour was more than enough to compensate for any extra protection provided by the seat belts.

There is a tendency to overestimate the risks of sudden imposed dangers where the consequences are severe, and to *underestimate* a risk if it has an effect in the long-term future, even if that effect is severe, for example, the health risks associated with smoking.

Q1.12 In a school of 1300 students, in one term 10 students contracted verrucas from the school pool. In a letter to parents the head teacher said there was a less than 1% chance of any child catching a verruca in any term. Was the figure she quoted correct and what assumptions had she made in making this statement?

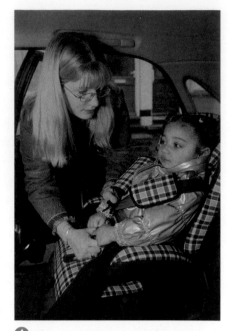

🔺 **Figure 1.23** Some research suggests that young children who wear rear seat belts are *more* likely to die in an accident than those who don't. This may be explained by parents' driving habits. Health risks are greatly affected by human behaviour.

Checkpoint ✓

1.6 List the circumstances that make people more likely to **a** underestimate and **b** overestimate the risk of an event happening. Suggest an example for each situation.

Different types of risk factor

In 2004, worldwide, there was one human heart attack every 4 seconds and a stroke every 5 seconds. In the UK the risk of any one of us having a heart attack in any one year is about 1 in 440 compared to 1 in 900 for a fatal stroke. However, these probabilities use figures for the whole population, giving averages which make the simplistic assumption that everyone has the same chance of having cardiovascular disease. This is obviously not the case.

The averages take no account of any risk factors – things that increase the chance of the harmful outcome. When assessing an individual's risk of bad health, all the contributing risk factors need to be established.

There are many different factors that contribute to health risks, for example:

- heredity
- physical environment
- social environment
- lifestyle and behaviour choices.

Heredity and risk

Some diseases, such as some simple genetic disorders, have a single risk factor. These diseases are 'determined' by the inheritance of a single defective gene and the risk of suffering from the conditions follow the Mendelian rules of inheritance that you have already met at GCSE.

For example, consider two people who are carriers of sickle cell anaemia. This is a genetic condition in which an abnormal form of haemoglobin, the oxygen-carrying protein found in red blood cells, is produced. This abnormal haemoglobin is less soluble and at low oxygen concentrations it crystallises, causing the red blood cells to become distorted. The sickle- or crescent-shaped cells are less efficient at carrying oxygen and can block small blood vessels.

These two people decide to try to have a child. The chances of this child, or any others they have, inheriting the defective form of the gene from both parents is 1 in 4, as shown in Figure 1.24.

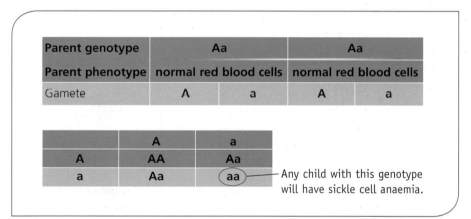

Parent genotype	Aa		Aa	
Parent phenotype	normal red blood cells		normal red blood cells	
Gamete	A	a	A	a

	A	a
A	AA	Aa
a	Aa	aa

Any child with this genotype will have sickle cell anaemia.

Figure 1.24 There is a 1 in 4 chance of any child inheriting two copies of the defective gene and suffering from sickle cell anaemia.

In some genetically inherited conditions the risks are very clear cut. Often though, inheritance is much more complex. Even in some conditions like sickle cell anaemia that are controlled by a single gene, it is now known that different mutations of that gene determine how severe the sufferer's condition is. This is true for cystic fibrosis, which you will study in detail in Topic 2.

Some diseases result from several genes interacting. In other diseases, genes have been identified that do not *cause* the condition in the sense that there isn't a clear-cut relationship between having the gene and having the condition. Rather, the genes *increase* the individual's susceptibility to the disease.

The chance that a person will suffer from cancer or cardiovascular disease, for example, is rarely the consequence of genetic inheritance alone. Many such diseases are multifactorial, with heredity, the physical environment, the social environment and lifestyle behaviour choices *all* contributing to the risk. The combination of risk factors experienced by the individual determines their risk of developing the disease.

Identifying risk factors – correlation and causation

To determine what the risk factors are for a particular disease, scientists who study disease look for **correlations** between potential risk factors and the occurrence of the disease.

Two variables are positively correlated when an increase in one is accompanied by an increase in the other (Figure 1.25A). For example, the length of a video and the percentage of the class asleep might be positively correlated. The number of cigarettes smoked over a lifetime and the chance of developing cardiovascular disease certainly are. If the values of one variable decrease while the other increases, there is a negative correlation (Figure 1.25B).

Large amounts of data are needed to ensure that the correlation is statistically significant, in other words not just an apparent correlation due to chance.

It is important to realise that a correlation between two variables does not necessarily mean that the variables are causally linked. Two variables are causally linked when a change in one is responsible for a change in the other. It is easy to think of variables that are correlated where there is no **causation**. For example, worldwide, speaking English as your first language correlates quite well with having a greater-than-average life expectancy. This, though, is simply because countries like the USA, UK, Australia and Canada have a higher-than-average standard of living and it is this that causes increased life expectancy through better nutrition, medical care and so on, rather than the language spoken.

It is because of this logical gap between correlation and causation that scientists try, whenever they can, to carry out experiments in which they can control variables, to see if altering one variable really does have the predicted effect. To do this scientists often set up a **null hypothesis**. They assume for the sake of argument that there will be no difference between an experimental group and a control group, and then test this hypothesis using statistical analysis.

Q1.13 Strong correlations have been reported between the following pairs of variables. In each case decide if there is likely to be a causal link between the variables or not. Suggest a possible reason for the correlation.
a shark attacks and ice cream sales
b children's foot sizes and their spelling abilities
c lung cancer and smoking

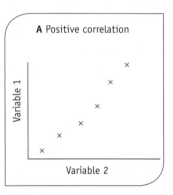

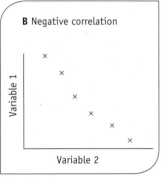

▲ **Figure 1.25 A** When an increase in one variable is accompanied by an increase in the other, there is a positive correlation, giving a scattergram rising from left to right. **B** With a negative correlation, one set of data increases while the other falls, resulting in a graph going down from left to right.

1.3 Risk factors for cardiovascular disease

Your chances of having coronary heart disease or a stroke are increased by several interrelated risk factors, the majority of which are common to both conditions. These include:

- genetic inheritance
- high blood pressure
- obesity
- blood cholesterol and other dietary factors
- smoking.

Some of these you can control, while others you can't.

There have been a number of large-scale studies to identify risk factors for coronary heart disease, including the Seven Countries Study and the Framlingham Study. The World Health Organisation MONICA study (MONItoring trends and determinants in CArdiovascular disease), involving over 7 million people in 21 countries over 10 years, confirmed earlier findings that linked several factors with increased incidence of the disease.

△ **Figure 1.26** Some of the potential risk factors for developing coronary heart disease are easy to identify, but may be difficult to control.

Do age and gender make a difference?

Q1.14 Look at Table 1.1. What happens to your risk of cardiovascular disease as you get older?

Q1.15 Does this mean that, at your age, you need not worry?

Q1.16 Do these data suggest that males and females face the same risk of cardiovascular disease?

Q1.17 Many people now think that, until menopause, a woman's reproductive hormones offer her protection from coronary heart disease. Do these data support this view? Is it valid to draw this conclusion from these data?

Activity

In **Activity 1.11** you compare data for coronary heart disease and stroke and look at trends over a 10-year period. **A1.11S**

The risk of premature death (before the age of 75) due to cardiovascular disease is higher for men than women in the UK. In both sexes, the prevalence of cardiovascular diseases increases with age. This may be due to the effects of ageing on the arteries; they tend to become less elastic and may be more easily damaged. With increasing age the risks associated with other factors may increase causing a rise in the number of cases of disease.

▷ **Table 1.1** Mortality data from diseases of the circulatory system for England and Wales in 2002. *Source: Office for National Statistics.*

Age/years	Female deaths	Male deaths
under 1	29	39
1–4	10	9
5–9	10	9
10–14	9	16
15–19	29	39
20–24	48	65
25–29	53	107
30–34	112	223
35–39	207	456
40–44	367	923
45–49	656	1 606
50–54	934	2 732
55–59	1515	4213
60–64	2512	5857
65–69	4476	8969
70–74	8714	13 700
75–79	15882	18 587
80–84	23036	19 512
85+	51072	22 749
Totals	**109 631**	**99 802**

My dad had a heart attack – will I?

If one or other of your parents suffer or suffered from cardiovascular disease, you are more likely to develop it yourself. There may be inherited predisposition for the disease.

However, the inheritance of cardiovascular disease is not a simple case of a single faulty gene for the condition being passed from one generation to the next. There are several genes that can affect your likelihood of developing cardiovascular disease and these interact with each other and the environment to produce an overall effect.

Families do not pass on just their genes. You may also acquire your parents' lifestyle and its associated risk factors such as smoking, lack of exercise and poor diet.

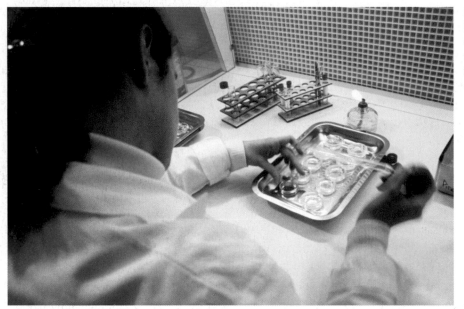

Activity

You can read about the role of genes in the sudden death of athletes in **Activity 1.12**. **A1.12S**

Figure 1.27 Genetic testing is becoming more common. Some companies now provide such testing so that individuals can see if they are at high or low risk of developing certain diseases.

Extension

The amount of cholesterol in your blood can also be affected by heredity. You can read more about inherited **hypercholesterolaemia** in **Extension 1.3**. **X1.03S**

High blood pressure

Elevated blood pressure, known as **hypertension**, is considered to be one of the most common factors in the development of cardiovascular disease. High blood pressure increases the likelihood of atherosclerosis occurring.

Blood pressure is a measure of the hydrostatic force of the blood against the walls of a blood vessel. You should remember that blood pressure is higher in arteries and capillaries than in veins. The pressure in an artery is highest during the phase of the cardiac cycle when the ventricles are contracting. This is the **systolic pressure**. Pressure is at its lowest in the artery when the ventricles are relaxed. This is the **diastolic pressure**.

Measuring blood pressure

A **sphygmomanometer** is a traditional device used to measure blood pressure. It consists of an inflatable cuff that is wrapped around the upper arm, and a manometer or gauge that measures pressure (Figure 1.28).

When the cuff is inflated the blood flow through the artery in the upper arm is stopped. As the pressure in the cuff is released the blood starts to flow through the artery. This flow of blood can be heard using a stethoscope positioned on the artery below the cuff. A pressure reading is taken when the blood first starts to spurt through the artery that has been closed. This is the *systolic* pressure. A second reading is taken when the pressure falls to the point where no sound can be heard in the artery. This is the *diastolic* pressure.

The SI units (International System of Units) for pressure are kilopascals, but in medical practice it is traditional to use millimetres of mercury, mmHg. (The numbers refer to the number of millimetres the pressure will raise a column of mercury.)

Blood pressure is reported as two numbers, one 'over' the other, for example $\frac{140}{85}$ (140 over 85). This means a systolic pressure of 140 mmHg and a diastolic pressure of 85 mmHg. For an average, healthy person you would expect a systolic pressure of between 100 and 140 mmHg and a diastolic pressure of between 60 and 90 mmHg.

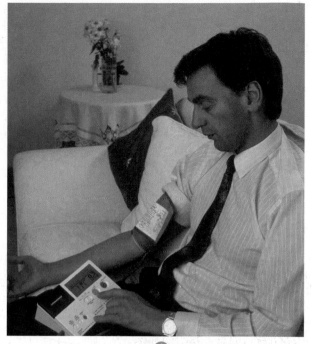

Figure 1.28 Nowadays blood-pressure monitors can give digital readouts.

systolic pressure, the maximum blood pressure when the heart contracts ⟶ $\frac{140}{85}$ ◄ diastolic pressure, the blood pressure when the heart is relaxed

Peter's blood pressure at age 42 was an incredible $\frac{240}{140}$.

What determines your blood pressure?

Contact between blood and the walls of the blood vessels causes friction, and this impedes the flow of blood. This is called peripheral resistance. The arterioles and capillaries offer a greater total surface area, resisting flow more, slowing the blood down and causing the blood pressure to fall. Notice in Figure 1.29 that the greatest drop in pressure occurs in the arterioles.

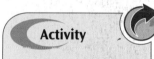

Activity

In **Activity 1.13** you use a sphygmomanometer, a blood-pressure monitor or the accompanying simulation to measure blood pressure. **A1.13S**

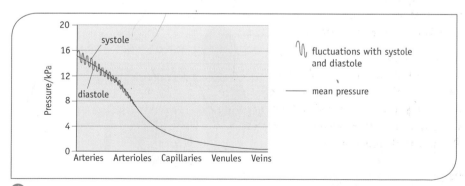

Figure 1.29 Blood pressure in the circulatory system. The fluctuations in pressure in the arteries are caused by contraction and relaxation of the heart. As blood is expelled form the heart, pressure is higher. During diastole, elastic recoil of the blood vessels maintains the pressure and keeps the blood flowing.

If the smooth muscles in the walls of an artery or an arteriole contract, the vessels constrict, increasing resistance. In turn, your blood pressure is raised. If the smooth muscles relax, the lumen is dilated, so peripheral resistance is reduced and blood pressure falls. Any factor that causes arteries or arterioles to constrict can lead to elevated blood pressure. Such factors include natural loss of elasticity with age, narrowing of arteries due to atherosclerosis, release of hormones such as adrenaline, or a high-salt diet. In turn high blood pressure can lead to atherosclerosis.

Tissue fluid formation and oedema

One sign of high blood pressure is **oedema**, fluid building up in tissues and causing swelling. Oedema may also be associated with kidney or liver disease, or with restricted body movement.

At the arterial end of a capillary, blood is under pressure. This forces fluid out through the capillary walls into the intercellular spaces, forming **tissue fluid** (also called interstitial fluid). Tissue fluid contains water and all the small molecules normally found in the plasma. The capillary walls prevent blood cells and larger plasma proteins from passing through, so these stay inside the capillaries.

Activity

Draw a **concept map** for blood pressure to bring together all the ideas covered. A pro-forma is available in **Activity 1.14** if you don't want to start from scratch. **A1.14S**

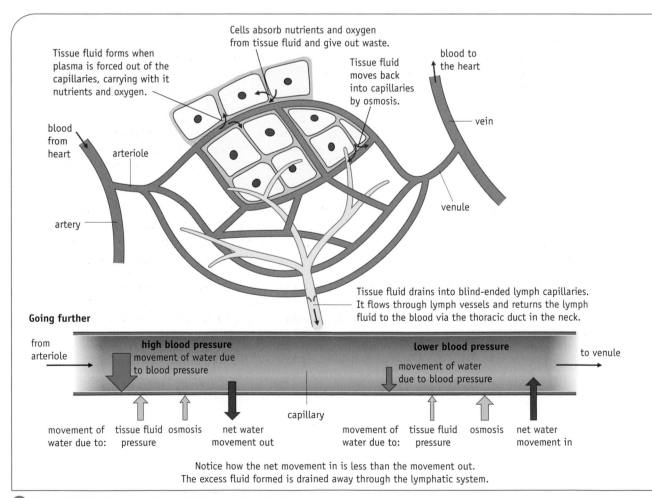

Figure 1.30 Production of tissue fluid in a capillary bed.

At the venous end of the capillary, blood pressure is lower so fluid is no longer being forced outwards. Due to the presence of blood cells and plasma proteins, and the loss of fluid, blood is more concentrated here (there is less water per unit volume) and therefore fluid moves back into the capillary by osmosis (see page 62). In addition, about 20% of the tissue fluid returns to the circulation via the **lymph vessels** (see Figure 1.30).

If blood pressure rises above normal, more fluid may be forced out of the capillaries. In such circumstances, fluid accumulates within the tissues causing oedema. During left-side heart failure (the most frequent type) there is an increase in pressure in the pulmonary vein and left atrium. This is because blood continues to flow out of the right side of the heart to the lungs and return to the heart due to the action of breathing muscles. There is a back-up of blood in the pulmonary capillaries, raising blood pressure. This causes tissue fluid to build up within the lungs, impairing gas exchange.

Q1.18 What would happen in the case of right-side heart failure?

More about tissue fluid formation

As plasma leaves the capillaries to form tissue fluid, dissolved substances such as oxygen and nutrients are carried out in the flow of fluid. This movement of fluid *out* of capillaries at the arteriole end and then back *in* to capillaries at the venule end depends on two separate forces, one the result of hydrostatic pressure and one the result of osmosis.

Hydrostatic pressure is the pressure exerted by a liquid, and this will be the same as 'blood pressure' in the vessels of the circulatory system. Blood pressure forces fluid out through the capillary walls. Tissue fluid bathing the cells also exerts hydrostatic pressure in the opposite direction, forcing water back into the capillaries.

Osmosis is the movement of water from a less concentrated solution (containing more water) to a more concentrated solution (containing less water). Blood will always be more concentrated than tissue fluid. This is due to the presence of plasma proteins, which are too large to pass out through the capillary walls with the other smaller, dissolved molecules. Osmosis will draw water into the capillaries from the tissues.

The net water movement will depend on the balance between the effects of hydrostatic pressure and osmosis (Figure 1.30). Fluid leaves at the arteriole end of a capillary bed and re-enters the capillaries at the venule end. The net water movement out of the capillary causing tissue fluid formation is slightly greater than the movement of water in the opposite direction at the venule end. In a normal situation the excess tissue fluid formed by this imbalance is drained away by the lymphatic system.

In the abnormal situation where excess tissue fluid formation results in oedema the movement of water out of the capillary due to blood pressure would be greater.

Dietary risk factors

Our choices of food, in particular the type and quantity of high-energy food, can either increase or decrease our risk of contracting certain diseases, including cardiovascular diseases.

Which nutrients store energy? **Carbohydrates**, **lipids** (often called fats and oils) and **proteins** are constituents of our food which store energy. Alcohol is a carbohydrate and can also provide energy. The relative energy content of these nutrients is shown in Table 1.2.

▼ **Table 1.2** Energy content of nutrients.

Nutrient	Energy available per gram/kJ
carbohydrates	16
lipids	37
proteins	17
alcohol	29

Energy units – avoiding confusion

Most packet foods these days detail the energy content per 100 g or other appropriate quantity. In Figure 1.31 notice the units used to express energy content for the chocolate bar. Why are two different units quoted? Which should we use?

Traditionally, energy was measured in **calories**; one calorie is the quantity of heat energy required to raise the temperature of 1 cm³ of water by 1°C. Food labels normally display units of 1000 calories, called **kilocalories** (also called **Calories** with a capital C).

The SI unit for energy is the joule (J), and 4.18 joules = 1 calorie. The **kilojoule** (1 kJ = 1000 joules) is used extensively in stating the energy contents of food. In the popular press the Calorie is used as the basic unit of energy, particularly with reference to weight control. Hence most food labels in the UK continue to quote both Calories and kilojoules. The chocolate bar shown in Figure 1.31 displays energy values in kcalories and kilojoules.

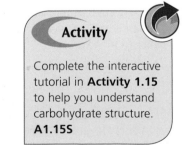

Nutrition Information		Per Bar	Per 100g
Energy	kJ	880	2260
	kcal	210	540
Protein	g	2.7	7.0
Carbohydrate	g	20.7	53.8
Fat	g	12.8	33.2

▲ **Figure 1.31** How much energy does this chocolate contain?

Q1.19 A newborn baby requires around 2000 kJ per day. Express this **a** in calories **b** in Calories.

Carbohydrates

The term carbohydrate was first used in the nineteenth century and means 'hydrated carbon'. If you look at each carbon in a carbohydrate molecule (see page 32), you should be able to work out why, bearing in mind that hydration means adding water.

Most people are familiar with sugar and starch being classified as carbohydrates, but the term covers a large group of compounds with the general formula $C_x(H_2O)_n$.

Q1.20 What are the ratios of carbon, hydrogen and oxygen in carbohydrates?

Sugars are either **monosaccharides**, single sugar units, or **disaccharides**, in which two single sugar units have combined in a condensation reaction.

> **Activity**
>
> Complete the interactive tutorial in **Activity 1.15** to help you understand carbohydrate structure. **A1.15S**

When a baby is born it survives entirely on milk (from the breast or a bottle) for a period that lasts from a couple of months to a year or more. Mother's milk is pretty much the same the world over. Aside from being a bit low in iron, fibre and some vitamins (e.g. vitamin C) it's a near perfect human diet.

Once weaning begins, a baby starts to take in solid food as well. This is where culture steps in. Adults across the globe eat very different foods and children, by and large, eat what they are given. The result is that we get used, as we grow up, to eating food that other people may think bizarre, even disgusting. Would you like to eat insect larvae, whale meat or sea cucumbers? Plenty of people do. Indeed, these foods are considered a delicacy in many countries.

There aren't just national differences in diet. In the UK, for example, there are regional differences – not just between England, Wales, Scotland and Northern Ireland but within each of these four areas too. Then there are differences related to social class (think about it!), age and gender. Marketing people know this only too well and carefully target food advertisements at the right 'segment' of the population.

And yet, one of the interesting things about what we eat nowadays is the result of globalisation. Nowadays most of us eat a greater variety of foods than our grandparents did. You probably eat Indian, Chinese and Italian foods, to mention just three nationalities. All in all we are remarkably adaptable as far as our diet goes.

See Figures 1.32 and 1.34. Chains of between 3 and 10 sugar units are known as **oligosaccharides** and long straight or branched chains of sugar units form **polysaccharides**.

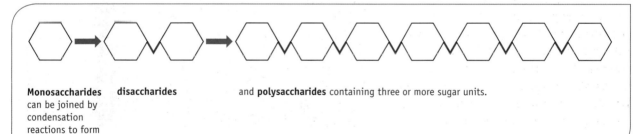

Monosaccharides can be joined by condensation reactions to form **disaccharides** and **polysaccharides** containing three or more sugar units.

Figure 1.32 A simplified diagram to show how simple sugar units (monomers) can be joined to form more complex carbohydrates (polymers).

Key biological principle: Large biological molecules are often built from simple subunits

Hydrogen, carbon, oxygen and nitrogen account for more than 99% of the atoms found in living organisms. Relatively simple molecules join together in different ways to produce many of the large important biological molecules.

Polymers, such as polysaccharides (Figure 1.32), proteins and nucleic acids, are made by linking identical or similar subunits, called **monomers**, to form straight or branched chains. Lipids are another group of biological molecules also constructed by joining smaller molecules together, though they are not polymers since they are not chains of monomers. Large biological molecules have structures that are well suited to their function.

In Topic 1 we are looking at the structure and function of some carbohydrates and lipids, returning in later topics to see how these molecules have many other roles. In Topic 2 the structure and function of nucleic acids and proteins will be examined in detail.

Monosaccharides

Monosaccharides are single sugar units with the general formula $(CH_2O)_n$, where n is the number of carbon atoms in the molecule. Monosaccharides have between three and seven carbon atoms, but the most common number is six. For example, the monosaccharides **glucose**, **galactose** and **fructose** all contain six carbon atoms and are known as **hexose** sugars.

A hexose sugar molecule has a ring structure formed by five carbons and an oxygen atom; the sixth carbon projects above or below the ring. The carbon atoms in the molecule are numbered, starting with 1 on the extreme right of the molecule. The side branches project above or below the ring, and their position determines the type of sugar molecule and its properties.

Support

To find out more about chemical reactions and bonding look at the Biochemistry support on the website.

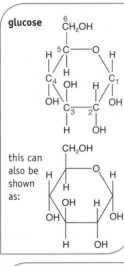

Glucose is important as the main sugar used by cells in respiration. Starch and glycogen are large complex carbohydrate molecules made up of glucose subunits joined together. When starch or glycogen is digested, glucose is produced. This can be absorbed and transported in the bloodstream to tissue cells.

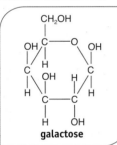

Galactose occurs in our diet mainly as part of the disaccharide sugar lactose, which is found in milk. Notice that the –OH groups on carbon 1 and carbon 4 lie on the opposite side of the ring compared with their position in glucose.

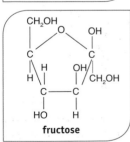

Fructose is a sugar which occurs naturally in fruit, honey and some vegetables. Its sweetness attracts animals to eat the fruits and so help with seed dispersal.

Monosaccharides provide a rapid source of energy. They are readily absorbed and require little or, in the case of glucose, no change before being used in cellular respiration. Glucose and fructose are found naturally in fruit, vegetables and honey; they are both used extensively in cakes, biscuits and other prepared foods.

Disaccharides

Two single sugar units can join together and form a disaccharide (double sugar) in a **condensation** reaction. A condensation reaction is so called because a water molecule is released as the two sugar molecules combine in the reaction. Condensation reactions are common in the formation of complex molecules. Figure 1.33 shows the formation of the disaccharide maltose by a condensation reaction between two glucose molecules. The bond that forms between the two glucoses is known as a glycosidic bond or link. The bond in maltose is known as a 1,4 glycosidic link because it forms between carbon 1 on one molecule and carbon 4 on the other.

Common disaccharides found in food are **sucrose**, **maltose** and **lactose**. Their structures are shown in Figure 1.34.

The white or brown crystalline sugar we use in cooking, and also golden syrup or molasses, is sucrose, extracted from sugar cane or sugar beet.

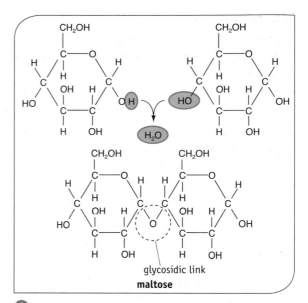

Figure 1.33 Two glucose molecules may join in a condensation reaction to form the disaccharide maltose. A water molecule is released during the reaction.

The glycosidic link between two sugar units in a disaccharide can be split by **hydrolysis**. This is the reverse of condensation: water is added to the bond and the molecule splits into two (Figure 1.35). Hydrolysis of carbohydrates takes place when carbohydrates are digested in the gut, and when carbohydrate stores in a cell are broken down to release sugars.

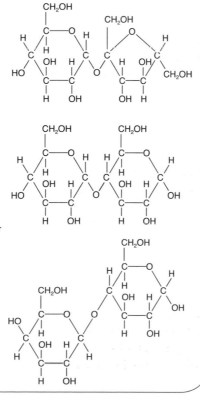

Sucrose
Sucrose, formed from glucose and fructose, is the usual form in which sugar is transported around the plant.

Maltose
Maltose, formed from two glucose molecules, is the dsaccharide produced when amylase breaks down starch. It is found in germinating seeds such as barley as they break down their starch stores to use for food.

Lactose
Glucose and galactose make up **lactose**, the sugar found in milk.

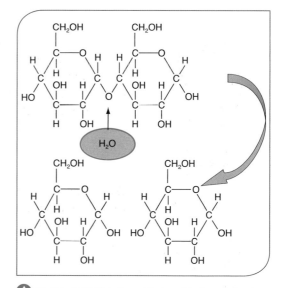

Figure 1.35 The glycosidic bond between the two glucose molecules in maltose can be split by hydrolysis. In this reaction water is added to the bond.

Figure 1.34 Disaccharides formed by joining two monosaccharide units.

Because polysaccharides and disaccharides have to be digested into monosaccharides before being absorbed, eating complex carbohydrates does not cause swings in blood sugar levels as does eating monosaccharides.

Lactose is the sugar present in milk. Many adults are intolerant of lactose and drinking milk will produce unpleasant digestive problems for these people. One solution is to hydrolyse the lactose in milk, which converts the *disaccharide* lactose into the *monosaccharides* glucose and galactose.

Industrially this is carried out using the enzyme lactase. Lactase can be immobilised in a gel, and milk is poured in a continuous stream through a column containing beads of the immobilised enzyme (Figure 1.36). Asian and Afro-Caribbean people have a particularly high rate of lactose intolerance, so the resulting lactose-free milk is particularly suitable for food-aid programs. Untreated milk would cause further problems for people already suffering from malnutrition and dehydration.

▲ **Figure 1.36** Whey waste from cheese-making contains lactose. Hydrolysis of the waste produces syrup which is used in the food industry.

> **Did you know?** Why do we have such a sweet tooth?
>
> We have taste receptors on the tongue for five main tastes – sweet, sour, bitter, salty and umami (the taste associated with monosodium glutamate, MSG). It is likely that the sweet-taste receptors enable animals to identify food that is easily digestible, whereas bitter-taste receptors provide a warning to avoid potential toxins. Humans, along with many other primates (apes and monkeys), have many more sweet-taste receptors than most other animals. Our sweet-taste receptors help us to identify when fruit is ready to eat.

> **Activity**
>
> In **Activity 1.16** you can immobilise lactase and use it to hydrolyse lactose. **A1.16S**

Oligosaccharides

Oligosaccharides form when between three and 10 sugar units join by condensation reactions. They are found in vegetables such as leeks, garlic, artichokes, lentils and beans. They are not easily digested and pass through the small intestine intact. Bacteria present in the large intestine then ferment them (use them in respiration), producing gases that cause flatulence.

Polysaccharides

Polysaccharides are polymers made up from simple sugars (monomers) joined by glycosidic links into long chains, as shown in Figure 1.37.

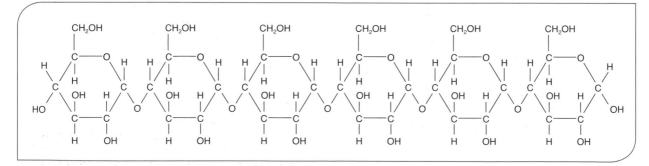

▲ **Figure 1.37** Glycosidic links join the glucose molecules that make up this polysaccharide.

There are three main types of polysaccharide found in food: starch and cellulose in plants, and glycogen in animals. Although all three are polymers of glucose molecules, they are sparingly soluble (they do not dissolve easily) and do not taste sweet.

Starch and glycogen act as energy storage molecules within cells. These polysaccharides are suitable for storage due to their low solubility in water. Starch or glycogen will not affect the water relations of a cell as they do not affect the concentration of water in the cytoplasm.

Starch, the storage carbohydrate found in plants, is made up of a mixture of two molecules, amylose and amylopectin. **Amylose** is composed of a straight chain of between 200 and 5000 glucose molecules with 1,4 glycosidic links between adjacent glucose molecules. The position of the bonds causes the chain to coil into a spiral shape. **Amylopectin** is also a polymer of glucose but it has side branches. A 1,6 glycosidic link holds the side branch onto the main chain. Figure 1.38 attempts to show these complex 3D structures. Starch grains in most plant species are composed of about 70–80% amylopectin and 20–30% amylose. The compact spiral structure of starch and its insoluble nature make it an excellent storage molecule. It does not diffuse across cell membranes and has very little osmotic effect within the cell.

Starch is a major source of energy in our diet, and common in many foods (Figure 1.39). It occurs naturally in fruit, vegetables and cereals, often in large amounts. The sticky gel formed when starch 'dissolves' in water makes it a good thickening agent and it is also added to many food products as a replacement for fat.

Bacteria and animals store **glycogen** instead of starch. Its numerous side branches (Figure 1.40) mean that it can be rapidly hydrolysed, giving easy access to stored energy. In humans glycogen is stored in the liver and muscles.

Cellulose in the diet is known as **dietary fibre**, and it is also referred to as a non-starch polysaccharide. Up to 10 000 glucose molecules are joined to form a straight chain with no branches (the glucose molecules are a slightly different structure to those found in starch). The structure of cellulose is considered in Topic 4; it is not required at this stage.

Indigestible in the human gut, cellulose has an important function in the movement of material through the digestive tract. Dietary fibre is thought to be important in the prevention of 'Western diseases' such as coronary heart disease, diabetes and bowel cancer.

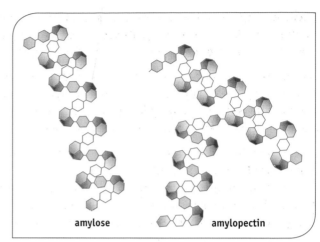

amylose   amylopectin

▲ **Figure 1.38** The two forms of starch: amylose and the branched chain amylopectin. The chains of glucose molecules coil to form a spiral. This is held in place by hydrogen bonds that form between the hydroxyl (OH) groups which project into the centre of the spiral.

▲ **Figure 1.39** Foods high in starch.

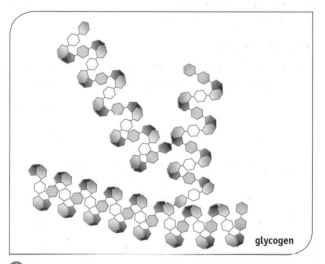

glycogen

▲ **Figure 1.40** Glycogen, the storage carbohydrate found in animal cells, has a branched structure similar to amylopectin.

Lipids

Lipids enhance the flavour and palatability of food, making it feel smoother and creamier (Figure 1.41). They supply over twice the energy of carbohydrates, 37 kJ of energy per gram of food, you will see how in Topic 7. This can be an advantage if large amounts of energy need to be consumed in a small mass of food.

Lipids are organic molecules found in every type of cell; they are insoluble in water but soluble in organic solvents such as ethanol. The most common lipids that we eat are **triglycerides**, used as energy stores in plants and animals. Triglycerides are made up of three **fatty acids** and one **glycerol** linked by condensation reactions (Figure 1.42).

△ **Figure 1.41** Which is more popular – with or without fat?

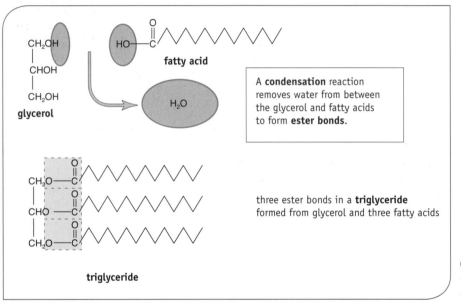

A **condensation** reaction removes water from between the glycerol and fatty acids to form **ester bonds**.

three ester bonds in a **triglyceride** formed from glycerol and three fatty acids

◁ **Figure 1.42** The formation of a triglyceride, a common type of lipid.

Saturated fats

If the fatty acid chains in a lipid contain the maximum number of hydrogen atoms, they are said to be **saturated**. In a saturated fatty acid the hydrocarbon chain is long and straight.

In a saturated fatty acid there are no carbon to carbon double bonds and no more hydrogens can be added to the chain. Animal fats from meat and diary products are major sources of saturated fats.

Straight, saturated hydrocarbon chains can pack together closely. The strong intermolecular bonds between triglycerides made up of saturated fatty acids result in fats that are solid at room temperature.

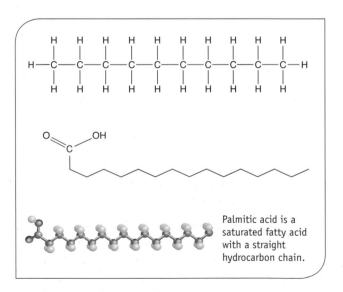

Palmitic acid is a saturated fatty acid with a straight hydrocarbon chain.

Unsaturated fats

Monounsaturated fats have one double bond between two of the carbon atoms in each fatty acid chain. **Polyunsaturated** fats have a larger number of double bonds. A double bond causes a kink in the hydrocarbon chain; these kinks prevent the unsaturated hydrocarbon chains packing closely together. The weaker intermolecular bonds between unsaturated triglycerides result in oils that are liquid at room temperature. Olive oil is particularly high in monounsaturated fats. Most other vegetable oils, nuts and fish are good sources of polyunsaturated fats.

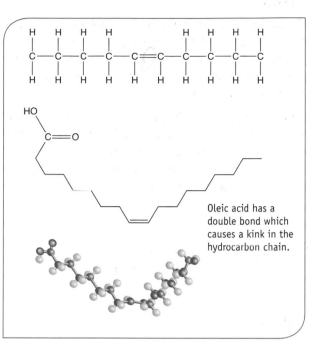

Oleic acid has a double bond which causes a kink in the hydrocarbon chain.

Other types of lipid

Cholesterol (Figure 1.43) is a short lipid molecule. It is a vital component of cell membranes with roles in their organisation and functioning. The steroid sex hormones (such as progesterone and testosterone) and some growth hormones are produced from cholesterol. Bile salts, involved in lipid digestion and assimilation, are formed from cholesterol. For all these reasons, cholesterol is essential for good health and is made in the liver from saturated fats. However, there are concerns that too high a blood cholesterol level can be bad for us. Cholesterol is found associated with saturated fats in foods such as eggs, meat and dairy products.

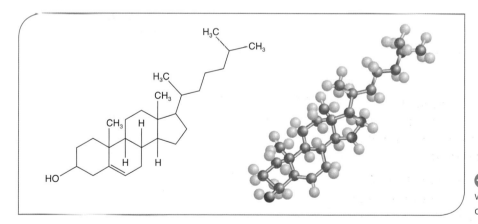

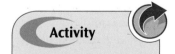

Figure 1.43 Alternative ways of showing the structure of cholesterol.

Phospholipids are similar to triglycerides but one of the fatty acids is replaced by a negatively charged phosphate group. Phospholipids are important components of cell membranes.

As well as supplying energy in the diet, fats also provide a source of **essential fatty acids**, that is, fatty acids that the body needs but cannot synthesise. In addition, fat-soluble vitamins (A, D, E and K) can only enter our diet dissolved in fats. Fats must therefore be present in a balanced diet to avoid deficiency symptoms. For example, a deficiency of linoleic acid (an essential fatty acid) can result in scaly skin, hair loss and slow wound healing.

Activity

Activity 1.17 Complete this interactive tutorial to understand lipid structure. **A1.17S**

The energy balance

Look on food labels and you often see recommended daily amounts for nutrients, along with daily energy requirements for men and women. But how much energy is right for each of us, and what happens if we don't get it right?

Getting it right

The UK Department of Health publishes dietary guidelines for most nutrients. They used to give recommended daily amounts but in 1991 these were largely replaced with dietary reference values (DRVs). These include:

- an estimated average requirement (EAR)
- a lower reference nutrient intake (LRNI)
- a higher reference nutrient intake (HRNI).

These effectively provide a range of values within which a healthy balanced diet should fall. Upper and lower limits have not been set for carbohydrates and fats. Instead, estimated average requirements are suggested plus the average percentage that should come from the different energy components of a diet. Tables 1.3 and 1.4 give the recommendations.

▼ **Table 1.3** Average daily energy requirements for adults in kJ per day.

Estimated average requirements for energy (EAR) for adults/kJ per day (kcalories per day)		
Age/years	Males	Females
19–50	10 600 (2550)	8100 (1940)

▼ **Table 1.4** Dietary guidelines for percentage of daily energy that should come from carbohydrate and fat.

% of daily total food energy intake excluding alcohol				
Fat		Carbohydrates		
Saturated	Unsaturated	Starch	Sugars	Non-starch polysaccharides
11	24	37	13	15

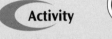

Activity

Activity 1.18 uses dietary analysis software for you to work out your energy budget and determine whether you are getting the right amount of energy and from the best sources.
A1.18S

Getting it wrong

We have to be aware that we need both carbohydrates and fats in our diet for good health, but that there are consequences if we get it wrong by consuming too much energy or if the percentage supplied by the various components differs greatly from the guidelines.

You need a constant supply of energy to maintain your essential body processes, such as the pumping of the heart, breathing and maintaining a constant body temperature. These processes go on all the time, even when you are completely 'at rest'. The energy needed for these essential processes is called the **basal metabolic rate** (BMR) and varies between individuals. BMR is higher in:

- males
- younger people
- heavier people
- more active people.

The 'average' person may require between 8000 and 10 000 kJ a day but an athlete may require double this quantity. Some cyclists consume four times this amount.

If you eat fewer kilojoules a day than you use, you have a negative energy balance and energy stored in the body will be used to meet the demand. A regular shortfall in energy intake will result in weight loss. If you routinely eat more energy than you use you have a positive energy balance. The additional energy will be stored and you will put on weight (Figure 1.44).

In the UK, it is estimated that around 60% of men and 40% of women are either overweight or obese. Approximately 20% of the population are obese, and this figure trebled from 1980 to 2000. The increasing prevalence of **obesity** among both children and adults has been called an epidemic in Western Europe and the USA.

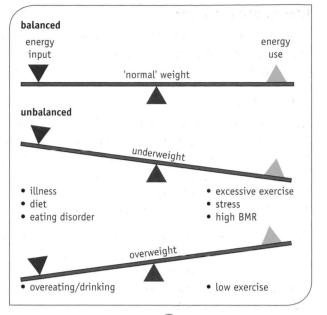

Figure 1.44 The balance between energy input and energy output determines whether the body maintains, gains or loses weight.

Defining 'overweight' and 'obese'

Body mass index (BMI) is an internationally accepted method of classifying body weight relative to a person's height. To calculate BMI, body mass (in kg) is divided by height (in metres) squared, i.e.

$$BMI = \frac{body\ mass/kg}{height^2/m^2}$$

For example, the BMI of a person with a body mass of 65 kg and height of 1.72 m is 22.0. This figure can then be used to identify the category of body weight to which that person belongs, as shown in Table 1.5.

Q1.21 Calculate the body mass index of a person with body mass of 85 kg and height 1.68 m. How would you describe the body weight of this person?

Table 1.5 The use of BMI to classify body weight.

BMI	Classification of body weight
<20	underweight
20–24.9	normal
25–29.9	overweight
30–40	obese
>40	severely obese

A high-fat diet will not necessarily result in weight gain if combined with a high level of physical activity. This was clearly illustrated by Ranulph Fiennes and Mike Stroud during their 1993 expedition to cross the Antarctica on foot – they found 23 000 kJ a day inadequate to meet their energy needs (Figure 1.45).

A poor diet, particularly one high in fat, and a sedentary lifestyle are the major contributing factors to the development of obesity. There is evidence in the UK that fat consumption has actually declined since 1990, but greater inactivity means that obesity and associated conditions are on the increase.

Consequences of obesity

Obesity increases your risk of coronary heart disease and stroke, even without other risk factors being present. The more excess fat you carry, especially around your middle, the greater the risk to your heart. Obesity raises your blood pressure and your blood cholesterol level, and can greatly increase your risk of type II diabetes. Type II diabetes is also referred to as non-insulin-dependent diabetes or late-onset diabetes. It, in turn, increases your risk of coronary heart disease and stroke.

Figure 1.45 Over 23 000 kJ a day and he still lost weight! Even with a padded jacket he looks thin.

In type II diabetes the body either does not produce sufficient insulin or the body fails to respond to the insulin that is produced. You probably know that **insulin** is the hormone that helps regulate blood glucose levels. After a meal, the level of blood glucose rises; in response to this change the **pancreas** produces insulin and secretes it into the bloodstream. The insulin causes cells to absorb glucose, thus returning the level in the blood to normal. Continually high levels of blood glucose due to frequent consumption of sugar-rich foods can reduce the sensitivity of cells to insulin, and type II diabetes results. It may take years to develop and may not even be diagnosed. It is thought that a million people in the UK are unaware that they have type II diabetes.

Obesity also raises your blood pressure and elevates your blood lipid levels, two classic risk factors for cardiovascular disease. Studies have shown a positive correlation between the percentage of saturated fat in the diet and high blood pressure. They have also shown a positive correlation between the percentage of saturated fat in the diet and increased incidence of cardiovascular disease. Much media attention, particularly in advertising, is focused on saturated fats and cholesterol.

Why is cholesterol such a problem?

There is a considerable amount of evidence to show that the higher your blood cholesterol level, the greater your risk of coronary heart disease (Figure 1.46).

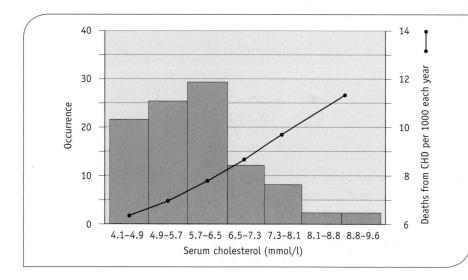

◀ **Figure 1.46** Blood cholesterol concentration related to coronary heart disease mortality in UK men aged 44–64 years. The joined up dots show the number of deaths from coronary heart disease with this serum cholesterol level per 1000 per year. The bars show the frequency of occurrence of the various serum cholesterol levels: for example, about 22% of UK men aged between 44 and 64 have a serum level of between 4.1 and 4.9 mmol per litre.

Q1.22 Comment on the relationship between serum cholesterol levels and the risk of death from coronary heart disease.

It is estimated that in the UK 45% of deaths from coronary heart disease in men and 47% of deaths from coronary heart disease in women are due to a raised blood cholesterol level (greater than 5.2 mmol per l). It is thought that 10% of deaths from coronary heart disease in the UK could be avoided if everyone had a blood cholesterol level of less than 6.5 mmol per l.

However, as you may realise, it's not quite as simple as that! Like all lipids, cholesterol is not soluble in water. In order to be transported in the bloodstream, insoluble cholesterol is combined with proteins to form soluble **lipoproteins**.

There are two major transport lipoproteins: **high-density lipoproteins** (**HDLs**) and **low-density lipoproteins** (**LDLs**). LDLs are the main cholesterol carrier in the blood. HDLs have a higher percentage of protein compared with LDLs, hence their higher density.

The triglycerides from saturated fats in our diet combine with cholesterol and protein to form LDLs. These circulate in the bloodstream and bind to receptor sites on cell membranes before being taken up by the cells. Excess LDLs in the diet overload these membrane receptors, resulting in high blood cholesterol levels. Saturated fats may also reduce the activity of LDL receptors so the LDLs are not removed from the blood, thus further increasing the blood cholesterol levels. This cholesterol will be deposited in the artery walls forming atheromas.

High-density lipoproteins are made when triglycerides from unsaturated fats combine with cholesterol and protein. HDLs transport cholesterol from the body tissues to the liver where it is broken down. This lowers blood cholesterol levels and helps remove the fatty plaques of atherosclerosis.

Monounsaturated fats are thought to help in the removal of LDLs from the blood. Polyunsaturated fats are thought to increase the activity of the LDL receptor sites so the LDLs are actively removed from the blood.

LDLs are associated with the formation of atherosclerotic plaques whereas HDLs reduce blood cholesterol deposition. Therefore, it is desirable to maintain a higher level of HDL (the so-called 'good cholesterol' or 'protective cholesterol') and a lower level of LDL (the so-called 'bad cholesterol'). Eating a low-fat diet which particularly avoids saturated fats will help reduce total blood cholesterol and also LDL cholesterol which constitutes the major component of the cholesterol risk for CVD.

Q1.23 Women generally have higher HDL:LDL ratios than men up until the menopause. What consequences would you expect this to have for the incidence of coronary heart disease in women related to men?

Q1.24 A person stops eating butter on their toast and starts using polyunsaturated spread instead. What effect will this have on their blood LDL levels? Explain your answer.

Measuring blood LDL level is difficult and expensive, but it is relatively easy to measure the total cholesterol level. Even home test kits are now available (Figure 1.47 on page 42). Testing for total cholesterol may be part of a cardiovascular assessment for someone suspected of being at risk of CVD.

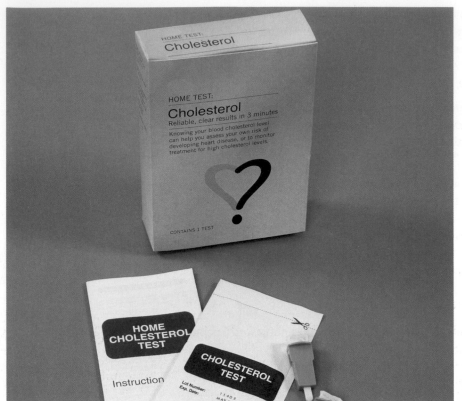

◀ **Figure 1.47** Cholesterol kits are available for home use.

Lifestyle risk factors

Smoking

Smoking cigarettes is one of the major risk factors for the development of cardiovascular disease. The constituents in smoke affect the circulation system in the following ways:

- The haemoglobin in red blood cells carries carbon monoxide from the smoke instead of oxygen. This reduces the supply of oxygen to cells. Any narrowing of the arteries due to atherosclerosis will reduce blood flow through the arteries in the heart and brain, further reducing the oxygen supply to the cells of the heart and brain. This will result in an increased heart rate as the body reacts to provide enough oxygen for the cells.
- Nicotine in smoke stimulates the production of the hormone adrenaline. This hormone causes an increase in heart rate and also causes arteries to constrict, both of which raise blood pressure.
- The numerous chemicals that are found in smoke can cause damage to the lining of the arterioles, triggering atherosclerosis.
- Smoking has also been linked with reduction in HDL cholesterol level.

Inactivity

The British Heart Foundation consider physical inactivity to be one of the most common risk factors for heart disease. They estimate that only three or four in every 10 men and two or three in every 10 women do sufficient exercise to give some protection against heart disease. It has been shown that being active can halve the risk of developing coronary heart disease.

Moderate exercise, such as walking, cycling or swimming, helps prevent high blood pressure and will lower blood pressure in someone suffering from hypertension. Exercise not only helps maintain a healthy weight, it also seems to raise HDL cholesterol without affecting LDL cholesterol levels. It also reduces the chance of developing type II diabetes and helps in the control of the condition.

A person who is physically active is much more likely to survive a heart attack or stroke compared with someone who has been inactive.

Other risk factors

The most important factors that increase the risk of cardiovascular disease are smoking, having high blood pressure, having too high a level of blood cholesterol and lack of physical activity. As we have seen, obesity is also a major risk factor. In addition, there are other things you should think about when deciding what to do if you want to lower your CVD risk.

The role of antioxidants

During reactions in the body, unstable radicals result when an atom has an unpaired electron, e.g. in the superoxide radical, $O_2 \bullet$. Radicals (sometimes known as free radicals) are highly reactive and can damage many cell components including enzymes and genetic material. This type of cellular damage has been implicated in the development of some types of cancer, heart disease and premature ageing. Some vitamins, including vitamin C, beta-carotene and vitamin E, can protect against radical damage. They provide hydrogen atoms that stabilise the radical by pairing up with its unpaired electron. The MONICA study (mentioned on page 25) found that high levels of antioxidants seemed to protect against heart disease. Partly because of their function as antioxidants, the current recommendation is to include five portions of fruit or vegetables per day in our diet.

Wine and some fruit juices contain chemicals which have antioxidant properties and also help stop platelets sticking together.

Q1.25 How might the antioxidants in wine help reduce the incidence of cardiovascular disease?

Salt

The Food Standards Agency recommended a salt intake of 6 g per day for an adult, but the UK average intake is double that figure. Eight per cent of the salt eaten comes from processed food. An average sized bowl of cereal can contain as much as 1 g of salt; a standard sized bag of crisps about 0.5 g. Notice on Figure 1.31 (page 30) how the value for sodium rather than salt is given; the value for salt is about 2½ times greater. A high-salt diet causes the kidneys to retain water. Higher fluid levels in the blood result in elevated blood pressure with the associated cardiovascular disease risks.

Activity

Activity 1.19 lets you determine if your diet contains enough antioxidant vitamins.
A1.19S

Stress

How you respond to stress in your life is very important. There is evidence that coronary heart disease is sometimes linked to poor stress management. In stressful situations the release of adrenaline causes arterioles to constrict, resulting in raised blood pressure. Stress can also lead to overeating, a poor diet and higher alcohol consumption which are all potential contributors to CVD.

Alcohol

Heavy drinkers are at far greater risk of heart disease and a number of other diseases. Heavy drinking raises blood pressure, contributes to obesity and can cause irregular heartbeat. There has been much research and debate concerning *potential* protective effects of moderate drinking.

If you have a glass of wine the alcohol it contains, 1 unit or 8 g, is very quickly absorbed, 20% through the wall of the stomach and the remainder through the walls of the small intestine. (One unit of alcohol is approximately the amount that an adult eliminates from the body in one hour.)

Excess alcohol consumption can result in direct tissue damage, including damage to the liver, brain and heart. Such damage contributes to an increased risk of cardiovascular disease. The liver has many functions but two of its main ones are processing carbohydrates, fats and proteins, and detoxification, including the removal and destruction of alcohol. High levels of alcohol can damage liver cells. This impairs the ability of the liver to remove glucose and lipids from the blood. In the liver alcohol is converted into ethanal, a three-carbon carbohydrate. Most of the ethanal is used in respiration but some ends up in very low-density lipoprotcins (LDLs), increasing the risk of plaque deposition.

Given these harmful consequences of alcohol, it seems remarkable to claim that moderate drinking may actually offer some degree of protection against cardiovascular disease. However, studies have shown a small protective effect of alcohol compared with abstinence. Moderate alcohol consumption is correlated with higher HDL cholesterol levels.

If you do drink, moderation is the key! The UK recommended limits to avoid health problems are 2–3 units per day for women, and 3–4 units per day for men, with no binge drinking. There is one unit of alcohol in half a pint of average strength beer, a glass of table wine or a measure of spirits.

Activity

Activity 1.20 is a teacher-led demonstration that lets you take part in an investigation of some factors that affect blood pressure and heart rate. **A1.20S**

Checkpoint ✓

1.7 Produce a concept map or table which shows the risk factors for CVD and their effects.

Activities

In **Activity 1.21** you can find out whether caffeine increases heart rate and blood pressure. **A1.21S**

In **Activity 1.22** you can review ideas about risk factors for CVD in this healthy heart quiz. **A1.22S**

1.4 Reducing the risks of cardiovascular disease

The risk of cardiovascular disease can be reduced in a range of different ways, including:

- stopping smoking
- maintaining blood pressure below 140/85 mmHg
- maintaining low blood cholesterol level
- maintaining a normal BMI
- taking more physical exercise
- moderate or no use of alcohol.

If people in the UK did not smoke, the British Heart Foundation estimates that 10 000 fewer men and women of working age would die from heart attacks each year. After stopping smoking, your risk of coronary heart disease is almost halved after only one year.

Controlling blood pressure

If a person is diagnosed with high blood pressure, changes in diet and lifestyle would be recommended. Medications are also available to reduce high blood pressure. One class of drugs, called diuretics, increase the volume of urine produced by the kidneys and thus rid the body of excess fluids and salt. This leads to a decrease in blood plasma volume and cardiac output (volume of blood expelled from the heart in a minute), which lowers blood pressure.

Sympathetic nerve inhibitors are another important class of drugs known as **antihypertensives** (drugs that reduce high blood pressure). Sympathetic nerves go from the brain to all parts of the body, including the arteries. When stimulated, they can cause the arteries to constrict, raising blood pressure. Antihypertensive drugs reduce blood pressure by *inhibiting* these nerves and so preventing them from constricting blood vessels.

Another class of drugs called beta-blockers reduce the heart rate and the heart's output of blood. They do this by blocking nerve receptors for hormones such as adrenaline, that would normally directly affect the sinoatrial node, increasing heart rate and cardiac output. By blocking the receptors for these hormones, beta-blockers reduce both heart rate and cardiac output. This helps reduce high blood pressure. Beta-blockers are often used to reduce the risk of someone having a second heart attack.

Reducing blood cholesterol levels

There is evidence from both the UK and the USA that untargeted general population cholesterol screening combined with dietary advice has little effect on lowering blood cholesterol levels. A further problem is associated with 'labelling' people. When told that they have high blood pressure, many people react by signing off sick!

Q1.26   Why do you think untargeted screening and dietary advice are not very effective at lowering blood cholesterol?

There is a need to reduce blood cholesterol in some people. One way to achieve this is through a low-fat diet. The media constantly remind us of the need to do this.

Table 1.6 shows the extent of cholesterol lowering that is obtained from following a low-fat diet in high-risk patients, namely people who had already experienced a heart attack, compared with the general population.

▼ **Table 1.6** The effect of lipid-lowering diets in reducing blood cholesterol levels.

	Blood cholesterol reduction/	
	mmol per l	%
general population	0.22	3
high risk patients	0.65	9

Q1.27 Can you suggest why the effect of dietary change appears to have been more successful in reducing blood cholesterol in patients who had experienced a heart attack than in the general population?

Individuals diagnosed with high cholesterol may also be prescribed cholesterol-lowering drugs. The main type used are statins. Statins work by inhibiting an enzyme involved in the production of LDL cholesterol.

A diet to reduce the risk of cardiovascular disease

A diet to offer protection against cardiovascular disease would include the following key features (Figure 1.48):

- Energy balanced.
- Reduced saturated fat.
- More polyunsaturated fats.
- Reduced cholesterol.
- Reduced salt.
- Increased soluble fibre. Soluble non-starch polysaccharides, such as pectins and guar gum, have been found to lower blood cholesterol. Soluble fibre is found in fruit, vegetables, beans, pulses and some grains (e.g. oats).
- Include oily fish. Fish such as mackerel, sardines, anchovies, salmon and trout contain *omega-3 fatty acids*, a group of polyunsaturated fatty acids with their first double bond between the third and fourth carbon atoms. These fatty acids are essential for cell functioning and have been linked to a reduction in heart disease and joint inflammation. The evidence for the importance of omega-3 fatty acids is seen in the Innuit in Greenland and the inhabitants of certain Japanese islands. They regularly eat oily fish and have very low rates of coronary heart disease.
- More fruit and vegetables. Fruits and vegetables contain antioxidants and often soluble fibre too.
- Include functional foods containing sterols and stanols. These are naturally produced substances in plants, similar to cholesterol. Stanols are saturated. Both sterols and stanols compete with cholesterol during its absorption in the intestine. Unfortunately we would have to eat vast quantities of foods such as vegetable oils and grains in order to reduce our cholesterol levels through this competition. Products have been developed that can incorporate sterols and stanols into everyday foodstuffs such as margarines (Figure 1.49). Controlled clinical trials of these functional foods show promising results: the amount of cholesterol absorbed from a food with stanols is about 20% compared with 50% absorption from a food without stanols.

▲ **Figure 1.48** A diet to reduce the risk of developing cardiovascular disease.

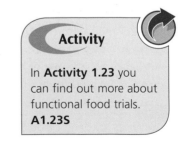

▲ **Figure 1.49** Some margarines and yoghurts include sterols or stannols to help lower cholesterol.

Activity

In **Activity 1.23** you can find out more about functional food trials. **A1.23S**

1.5 Treatments for cardiovascular disease

The NHS National Service Framework for coronary heart disease has set out standards and services that should be available throughout England for people with diagnosed coronary heart disease. These include:

- advice about how to stop smoking (including nicotine replacement therapy)
- information about other modifiable risk factors
- advice and treatment to maintain blood pressure below 140/85 mmHg
- low dose aspirin (75 mg daily)
- dietary advice to lower serum cholesterol concentrations to either below 5 mmol/l or by 30% (whichever is greater)
- beta-blockers for people who have had a myocardial infarction
- warfarin or aspirin for people over 60 years old who have atrial fibrillation
- very careful control of blood pressure and blood glucose in people who have diabetes.

If someone has had a heart attack or stroke, or is identified as being at high risk of one, in addition to the lifestyle changes and drug treatments to reduce blood pressure and blood cholesterol they may be given drugs to prevent formation of a blood clot in the artery.

Anticoagulant and platelet inhibitory drugs

The tendency of blood to form clots is reduced by platelet inhibitory drugs and anticoagulant drugs. Aspirin is an anti-platelet drug; it reduces the 'stickiness' of platelets and the likelihood of clot formation. A combined treatment involving a daily dose of aspirin and a second anti-clotting drug called clopidogrel has a dramatic effect. Trials in 2001 involving 12 500 patients showed that the combination reduces the risk of a patient dying from heart disease, having a repeat heart attack or suffering a stroke by 20–24%. Another anticoagulant drug is warfarin. This can be taken orally for extended periods of time to prevent clotting.

In an emergency situation when an artery is blocked, a clot-busting drug may be administered. Streptokinase is frequently used. Injected into the patient, the enzyme circulates within the blood and breaks down the clot.

Other approaches

Research into gene therapy suggests that injecting genes which code for a protein that enhances blood vessel growth may prove successful in helping to relieve some patients' symptoms.

All in all, this is a truly exciting area of biomedical science and one in which new developments seem to occur daily. Try to keep abreast of these developments throughout your biology course, whether you are taking the one-year AS course or the two-year AS and A2 course.

Extension

You can read more about new treatments for coronary heart disease in **Extension 1.4**. 'New treatments for cardiovascular disease' provides a fine start. It even gives you the opportunity to observe surgery! **X1.04S**

Did you know? Heart surgery

Those with severe coronary heart disease (who have had a heart attack like Peter) may need surgery. Coronary angioplasty, also known as balloon angioplasty, uses a catheter which is inserted into an artery in the groin and guided by X-ray imaging up to the narrowed coronary artery. A tiny balloon at the tip of the catheter is inflated and deflated to stretch or open the constriction and improve the passage for blood flow (Figure 1.50). The balloon-tipped catheter is then removed. The patient remains awake throughout the procedure.

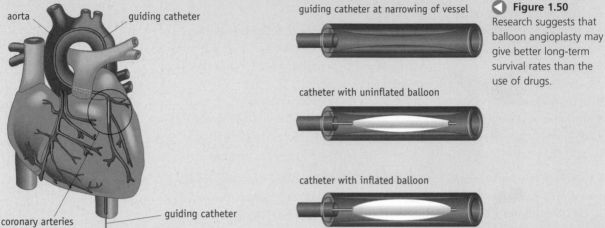

guiding catheter at narrowing of vessel

catheter with uninflated balloon

catheter with inflated balloon

◀ **Figure 1.50**
Research suggests that balloon angioplasty may give better long-term survival rates than the use of drugs.

aorta
guiding catheter
coronary arteries
guiding catheter

In a coronary artery bypass operation, a blood vessel, usually taken from the leg or chest, is grafted onto the blocked artery, bypassing the blocked area. Two, three or four blocked arteries can be bypassed at once – a double, triple or quadruple bypass (Figure 1.51). The blood can then go around the obstruction to supply the heart with enough blood to relieve chest pain. Peter had a quadruple bypass.

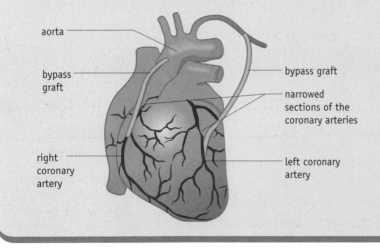

aorta
bypass graft
bypass graft
narrowed sections of the coronary arteries
right coronary artery
left coronary artery

◀ **Figure 1.51** A vein taken from the patient's leg is used to bypass the sections of coronary artery that are narrowed. The photograph above shows a section of vein that has been grafted onto a patient's heart.

Think back to Mark and Peter. What is most surprising is that Mark was only 15 when he had his stroke. Is it likely that Mark suffered from atherosclerosis? Mark had no obvious risk factors that would have alerted him to the possibility of having a stroke. He reports having taken exercise, eaten a reasonably healthy diet and having not smoked. If you read Mark's full story in Activity 1.1 at the start of the topic you will find that he had a type of stroke in which a blood vessel supplying blood to the brain bursts.

Did you know? Haemorrhagic stroke

Blood vessels on the surface of, or within the brain are susceptable to bursting, resulting in a stroke. A haemorrhagic stroke occurs when a blood vessel supplying blood to the brain bursts. If the burst occurs within the brain it is known as an *intracerebral haemorrhage*, whereas bursting of a vessel on the surface causes what is known as a *subarachnoid* *haemorrhage*. Look at Figure 1.52 and work out why the different types of strokes were given these names. There is no need for you to remember the names but it is worth being aware of them as doctors and medical scientists use this sort of language.

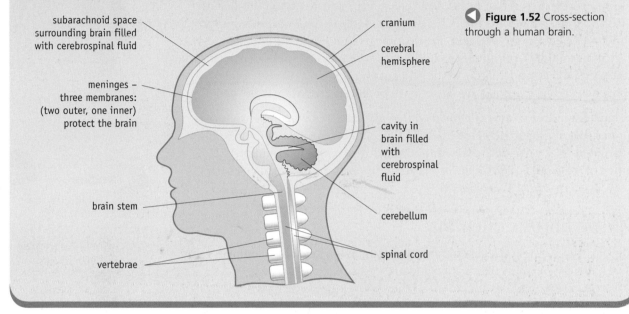

Figure 1.52 Cross-section through a human brain.

subarachnoid space surrounding brain filled with cerebrospinal fluid

cranium

cerebral hemisphere

meninges – three membranes: (two outer, one inner) protect the brain

cavity in brain filled with cerebrospinal fluid

brain stem

cerebellum

vertebrae

spinal cord

An artery can burst due to an aneurysm, where blood builds up behind a section of artery that has narrowed as a result of atherosclerosis. However, with no risk factors and in one so young the likelihood of atheroma deposits having built up to this extent in Mark's arteries seems unlikely. There was no history of stroke in Mark's family so he thinks that he inherited a gene causing him to have thin artery walls more prone to bursting.

Q1.28 What additional information would you need from Mark to determine if his stroke was due to this type of inherited condition or due to atherosclerosis?

Mark remains healthy today (March 2005) and has not had a recurrence of the problem.

Peter is lucky to be alive having had a blood pressure of 240/140 mmHg, two heart attacks and heart surgery. If you look back at his story you will recall that his father died aged 53 from a heart attack. This suggests that there may have been an inherited predisposition for the condition. Thankfully, Peter's active lifestyle will have helped him survive.

Q1.29 Peter has two daughters and a son. What advice would you recommend he give them to reduce their chances of developing CVD?

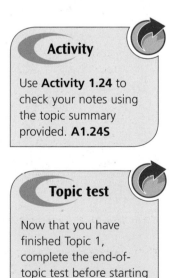

Activity

Use **Activity 1.24** to check your notes using the topic summary provided. **A1.24S**

Topic test

Now that you have finished Topic 1, complete the end-of-topic test before starting Topic 2.

Topic ② Genes and health

Why a topic called Genes and health?

It is now recognised that our genes have a major part to play in our health. Nearly 10 000 genetic disorders have been described. Some are very minor and create little or no problem; for example, colour blindness and hairy ears are both inherited conditions. Others have more serious consequences for the affected individual. Albinism, cystic fibrosis, Huntington's disease, sickle cell anaemia and haemophilia are all serious conditions caused by faulty genes.

Thankfully, most serious genetic disorders are rare. But when one does occur it has a significant effect on the person's life and on the lives of their family members. Not only must affected individuals deal with the condition itself, they may also face difficult decisions about the possible inheritance of the condition by the next generation.

Cystic fibrosis dilemma

In this topic we follow Claire and Nathan who face a daunting dilemma – should they have a child that could have cystic fibrosis (Figure 2.1)? Claire's mother had the condition, so they know it might be passed on to

Weblink

To find out the life expectancy for someone with cystic fibrosis today visit the Cystic Fibrosis Trust website.

▲ **Figure 2.1** Will this be the outcome for Claire and Nathan? Will they decide to have a baby? Will their baby have CF?

any child they might have. To make the decision they need more information. If the child does inherit cystic fibrosis how will it affect the child's life and theirs? How is cystic fibrosis inherited? What are the chances of any child they have inheriting the condition? Can genetic screening help? What treatments are available now and what might be possible in the future? These are just some of the questions that need answers if Nathan and Claire are to make an informed choice.

Cystic fibrosis (CF) is one of the most common genetic diseases, affecting approximately 7500 people in the UK. One in 25 of the population carries the faulty CF allele. Every week five babies are born with CF and three young people die from cystic fibrosis, usually as a result of lung damage. In the 1960s the average life expectancy for a child with CF was just 5 years. By the year 2000 life expectancy had risen to 31.

Claire is well aware of the outward symptoms of CF; she saw the problems her mother Valerie had with breathing, including a troublesome cough and repeated chest infections. Valerie also had to be very careful about her diet to ensure she overcame problems with digestion and maintained her weight. But Claire and Nathan want to know in more detail what causes these outward symptoms. The symptoms of CF concern a sticky mucus layer that lines many of the tubes and ducts in the gas exchange, digestive and reproductive systems.

The symptoms of cystic fibrosis and the way it is inherited have been known for a long time – it was first accurately described by Dorothy Andersen, a New York pathologist, in 1938. However, the genetic cause of the disease was not identified until 1989, by a group in Toronto led by Lap-Chee Tsui.

Review

Are you ready to tackle Topic 2 *Genes and health*?

Complete the GCSE review and GCSE review test before you start.

Overview of the biological principles covered in this topic

In this topic you will study how changes in DNA can result in genetic disease using the example of cystic fibrosis. You will first look in detail at the symptoms of cystic fibrosis, extending your previous knowledge of the structure of the lungs and of gas exchange to see the importance of surface area to volume ratios in biology.

To understand the symptoms of cystic fibrosis you will study cell membrane structure and how substances move across membranes. To explain how faults arise in the cell surface membranes of a person with CF you will gain detailed knowledge of how genes code for proteins, how proteins are made and how their function is dependent on their structure. You will find out how genes play their role in inheritance.

You will discover how cystic fibrosis is treated conventionally and how genetic screening and gene therapy may be used to help people with cystic fibrosis. Throughout the topic we will consider the ethical issues raised by these new technologies and you will learn how ethical arguments can be evaluated.

Activities

Activity 2.1 will give you an overview of cystic fibrosis, and let you find out more about Claire's family and the problems she and Nathan could face in the future. **A2.01S**

Read the stories in **Activity 2.2** to see how cystic fibrosis has affected some people, and how they and their families cope. **A2.02S**

2.1 The effects of CF on the lungs

The role of mucus in the lungs

The lungs allow efficient gas exchange between the atmosphere and blood. Air is drawn into the lungs via the trachea due to low pressure in the lungs, created by the movement of the ribs and diaphragm. The trachea divides into two bronchi which carry air to each lung. Within each lung there is a tree-like system of tubes ending in narrow tubes, bronchioles, attached to tiny balloon-like alveoli (Figure 2.2A). The alveoli are the main sites of gas exchange.

There is nothing unusual about having a layer of **mucus** in the tubes of the gas exchange system. Everyone normally has a thin coating of mucus in these tubes, produced continuously from goblet cells in the walls of the airways (Figure 2.2B). Any dust, debris or microorganisms that enter the airways become trapped in the mucus. This is continually removed by the wave-like beating of **cilia** that cover the **epithelial** cells lining the tubes of the gas exchange system. People with CF, like Valerie, have mucus that is drier than usual resulting in a sticky mucus layer that the cilia find difficult to move (Figure 2.2B).

This sticky mucus in the lungs has two major effects on health. It increases the chances of lung infection and it makes gas exchange less efficient, particularly in the later stages of the disease.

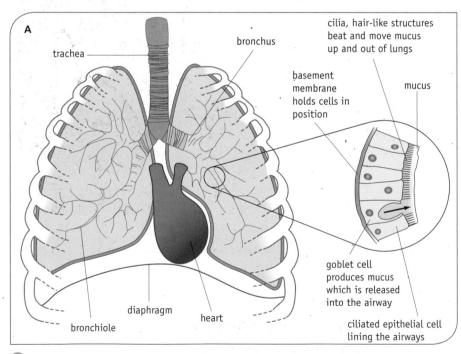
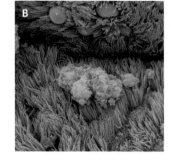

🔺 **Figure 2.2 A** The fine structure of the lungs. Cilia, hair-like extensions from the surface of cells lining the trachea, bronchi and bronchioles, have an important role in keeping the lungs clean. By a repeated beating motion they move mucus and particles up and out of the lungs. **B** Scanning electron micrograph of a clump of mucus (green/orange) surrounded by cilia (purple) in the lungs of someone with CF. Rounded goblet cells (light blue) secrete the mucus.

Did you know? What are epithelial cells?

Epithelial cells form the outer surface of many animals including mammals. They also line the cavities and tubes within the animal, and cover the surfaces of internal organs. The cells work together as a tissue known as **epithelium**.

The epithelium may consist of one or more layers of cells sitting on a **basement membrane**. This is made of protein fibres in a jelly-like matrix. There are several different types of epithelia.

The walls of the alveoli and capillaries are squamous or pavement epithelium. The very thin flattened cells fit together like crazy paving. The cells can be less than 0.2 μm thick.

In the small intestine the epithelial cells extend out from the basement membrane. The column-shaped cells make up columnar epithelium. The free surface facing the intestine lumen is normally covered in **microvilli** to increase surface area.

In the trachea, bronchi and bronchioles there are ciliated epithelial cells with cilia (hair-like structures) on the free surface. These cilia beat and move substances along the tube they line. The ciliated columnar epithelium of the gas exchange airways appears to be stratified (composed of several layers), but in fact each cell is in contact with the basement membrane. It appears to have several layers because some cells have their nucleus at the base of the cell while in others it is in the centre, giving the impression of different layers. This epithelium is therefore known as pseudostratified.

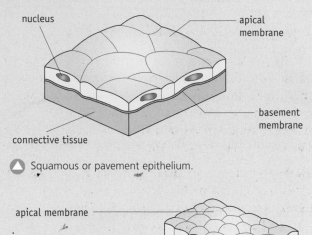

▲ Squamous or pavement epithelium.

▲ Columnar epithelium.

▲ Ciliated epithelium. Magnification ×5400.

CF problems

How sticky mucus increases the chances of lung infections

Microorganisms become trapped in the mucus in the lungs. Some of these can cause illness – they are **pathogens**. The mucus is normally moved by cilia into the back of the mouth cavity where it is either coughed out or swallowed, thus reducing the risk of infection. Acid in the stomach kills any microorganisms that are swallowed.

With CF the mucus layer is so sticky that cilia cannot move the mucus. Mucus production still continues, as it would in a normal lung, and the airways build up layers of thickened mucus. There are low levels of oxygen in the mucus, partly because oxygen diffuses slowly through it, and partly because the epithelial cells use up more oxygen in CF patients. Unfortunately dangerous bacteria thrive in anaerobic conditions.

Q2.1 Why might failure to move mucus create a problem?

Q2.2 Why does swallowing mucus reduce the risk of infection?

Repeated infections can eventually weaken the body's ability to fight the pathogens, and cause damage to the structures of the gas exchange system. White blood cells fight the infections within the mucus but as they die they break down, releasing DNA which the makes mucus even stickier.

How sticky mucus makes gas exchange less efficient

Gases such as oxygen cross the walls of the alveoli into the blood system by diffusion. To supply enough oxygen to all the body's respiring cells gas exchange must be efficient. The fine structure of the lungs helps to maximise this.

Key biological principle: The effect of increase in size on surface area

Living organisms have to exchange substances with their surroundings. For example, they take in oxygen and nutrients and get rid of waste materials such as carbon dioxide. In unicellular organisms the whole cell surface membrane is the exchange surface. Oxygen, for example, diffuses into a cell down a **concentration gradient** (from a high to a low concentration). A gradient is maintained by the cell using the substances it absorbs. Oxygen is used continually for respiration, for example.

The larger an organism, the more exchange needs to take place to meet the organism's needs. Larger multicellular organisms have more problems absorbing substances because of the size of the organism's surface area compared with its volume. This is known as the **surface area to volume ratio**, calculated by dividing the total surface area by the volume.

Q2.3 For each of the 'organisms' below work out its surface area, volume and then surface area to volume ratio.

Q2.4 As the organism grows larger what happens, quantitatively, to **a** its surface area **b** its volume **c** its surface area to volume ratio?

Q2.5 Assuming that this organism relies on diffusion across its outer surface for exchange, why would it have problems if it grew any larger?

Q2.6 If you compared a tiger, a horse and a hippopotamus, which would have the smallest surface area to volume ratio?

Activity

Complete **Activity 2.3** to investigate the effect of surface area to volume ratio on uptake by diffusion. **A2.03S**

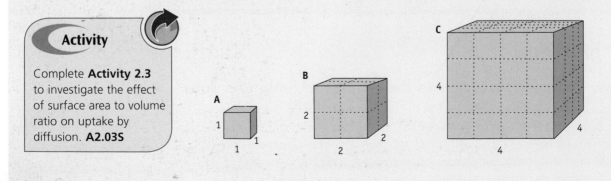

It is clear that as organisms get larger, the surface area per unit of volume gets less. If larger organisms relied on their general surface area for uptake of substances they could not absorb enough to survive.

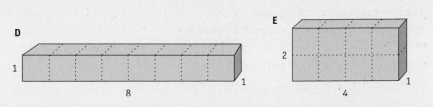

How can an organism increase in volume while still managing to absorb enough nutrients by diffusion?

Q2.7 Work out the surface area, volume and surface area to volume ratio of organisms D and E.

Q2.8 Look at the values you have calculated for B, D and E. What do you notice?

Q2.9 If a slug, an earthworm and a tapeworm all had the same volume, which would have the largest surface area to volume ratio?

All three blocks B, D and E have the same volume but they have very different surface areas. The most elongated block, D, has the largest surface area to volume ratio.

Relying on the outer body surface for gas exchange is only possible in organisms with a very small volume, or in larger organisms that have a high enough surface area to volume ratio such as worms with a tubular or flattened shape.

Q2.10 Why would a land-living organism not be likely to use its entire external surface covering for gas exchange? Think of possible problems this could cause, apart from surface area to volume ratio.

Larger organisms have a variety of special organs that increase the surface area for exchange, thus increasing the surface area to volume ratio. For example, our lungs provide a large surface for gas exchange while minimising heat and water loss from the moist surface. Our digestive system provides a large surface for food absorption.

Q2.11 Which of the four organisms below, all with the same volume, has the largest surface area for exchange?

Q2.12 Name two organs that have developed to aid exchange.

Q2.13 In humans, how do the substances that are absorbed get to all the distant parts of the body?

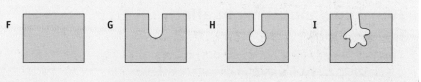

Gas exchange surfaces

Within the lungs alveoli provide a large surface area for exchange of gases between the air and the blood. Look at Figure 2.3 and identify features of the **gas exchange surface** that you think would ensure quick and efficient exchange between air in the alveoli and the blood.

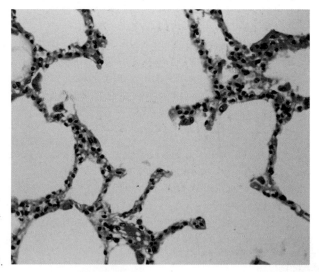

▶ **Figure 2.3** Ventilation of the lungs ensures that the air in the alveoli is frequently refreshed. This helps maintain a steep concentration gradient and maximise gas exchange across the walls of the alveoli.

You should have noticed some of these features of the gas exchange surface:

- large surface area of the alveoli
- numerous capillaries around the alveoli
- thin walls of the alveoli and capillaries making a short distance between the alveolar air and blood in the capillaries.

The body's demand for oxygen is enormous, so diffusion across the alveolar wall needs to be rapid. The rate of diffusion is dependent on three properties of gas exchange surfaces:

- **Surface area** – rate of diffusion is directly proportional to the surface area. As the surface area increases the rate of diffusion increases.
- **Concentration gradient** – rate of diffusion is directly proportional to the difference in concentration across the gas exchange surface. The greater the concentration gradient the faster the diffusion.
- **Thickness of the gas exchange surface** – rate of diffusion is inversely proportional to the thickness of the gas exchange surface. The thicker the surface the slower the diffusion.

$$\text{rate of diffusion} \propto \frac{\text{surface area} \times \text{difference in concentration}}{\text{thickness of the gas exchange surface}}$$

This is known as **Fick's law**.

The large surface area of the alveoli, the steep concentration gradient between the alveolar air and the blood (maintained by ventilation of the alveoli) and the thin alveolus and capillary walls combine to ensure rapid diffusion across the gas exchange surface (Figure 2.4).

How sticky mucus might affect gas exchange

The sticky mucus layer in the bronchioles of a person with cystic fibrosis tends to block these narrow airways, preventing ventilation of the alveoli below the blockage. This reduces the number of alveoli providing surface area for gas exchange. Blockages are more likely at the narrow ends of the airways. These blockages will often allow air to pass when the person breathes in but not when they breathe out, resulting in over-inflation of the lung tissue beyond the block. This can damage the elasticity of the lungs.

People with CF find it difficult to take part in physical exercise because their gas exchange system cannot deliver enough oxygen to their muscle cells. The oxygen is needed in the chemical processes of aerobic respiration, which release the energy used to drive the contraction of the muscles during exercise. People with CF become short of breath when taking exercise but exercise is very beneficial to them.

Activities

In **Activity 2.4** you will examine slides of the alveoli to observe the features that aid diffusion into the bloodstream. **A2.04S**

In the web tutorial, **Activity 2.5**, you can investigate the surface area of the lungs. **A2.05S**

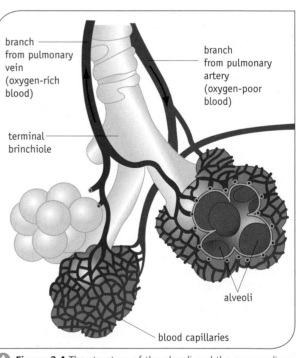

branch from pulmonary vein (oxygen-rich blood)

branch from pulmonary artery (oxygen-poor blood)

terminal brinchiole

alveoli

blood capillaries

▲ **Figure 2.4** The structure of the alveoli and the surrounding capillaries ensures that there is rapid diffusion across the gas exchange surface.

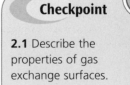

Checkpoint

2.1 Describe the properties of gas exchange surfaces.

2.2 Why is CF mucus so sticky?

In people with CF, the mucus layer on the surface of the epithelial cells is sticky because it contains less water than normal. The reduced water level is due to abnormal salt and water transport across the cell surface membranes of cells lining the airways. This is the result of a faulty transport protein channel in the membrane. To understand what is going on you need to be clear about the structure and function of proteins, the structure of cell surface membranes and how substances are transported in and out of cells. You also need to know how proteins are made so you can understand what has gone wrong with the protein channel.

Key biological principle: Protein structure is the key to protein function

Proteins have a wide range of functions in living things. Hormones, antibodies and enzymes are all protein molecules. Various proteins make up muscles, ligaments, tendons and hair. Proteins are also components of cell membranes and have important functions within the membrane. All proteins are composed of the same basic units: amino acids. There are 20 different amino acids that occur commonly in proteins. Plants can make all these amino acids whereas animals have to obtain some through their diet – these amino acids are known as essential amino acids.

Figure 2.5 shows the basic structure of an amino acid. All amino acids contain an amine group $-NH_2$, and a carboxylic acid group $-COOH$, attached to a central carbon atom. Each type of amino acid has a different side chain, called a residual or R group. Glycine is the simplest amino acid with a hydrogen, $-H$, forming its R group. Alanine's R group is a CH_3 group. Some amino acids have R groups with more complex carbon ring structures.

Primary structure

Two amino acids join in a condensation reaction to form a dipeptide, with a peptide bond forming between the two subunits, as shown in Figure 2.6. This process can be repeated to form polypeptide chains which may contain thousands of amino acids. A protein is made up of one or more of these polypeptide chains. The sequence of amino acids in the polypeptide chains is known as the **primary structure** of a protein.

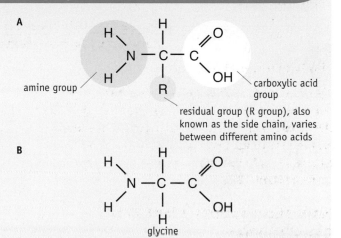

▲ **Figure 2.5 A** The general structure of amino acids. **B** The structure of glycine.

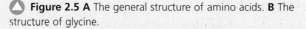

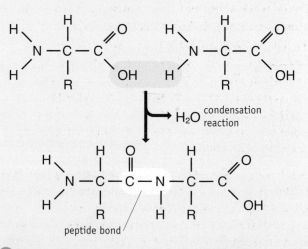

▲ **Figure 2.6** Two amino acids link in a condensation reaction to form a dipeptide. Numerous amino acids link together in this way to form a polypeptide.

Secondary, tertiary and quaternary structures

Interactions between the R groups of the amino acids in the chain cause the chain to twist and fold into a three-dimensional shape. Lengths of the chain may first coil into α-helices or β-pleated sheets (Figure 2.7). These are known as the secondary structure. The chain then folds into its final three-dimensional shape. The shape of the folded protein, the tertiary structure, depends on the interactions between the side branches of the amino acids that make up the chain (Figure 2.8).

Secondary structure

The chain of amino acids may twist to form an α-helix, a shape like an extended spring (Figure 2.7A). Within the helix, hydrogen bonds form between the C=O of the carboxylic acid and the –NH of the amine group of different amino acids, stabilising the shape.

Several chains may link together, with hydrogen bonds holding the parallel chains together in an arrangement known as a β-pleated sheet (Figure 2.7B). Within one molecule there may be sections with α-helices and other sections that contain β-pleated sheets.

Support

To find out more about hydrogen bonding have a look at the Biochemistry support on the website.

▼ **Figure 2.7** The secondary structure of proteins:
A A tightly wound α-helix.
B A β-pleated sheet.

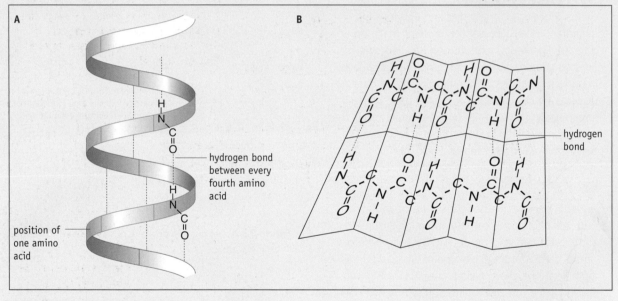

Tertiary and quaternary structure

A polypeptide chain often bends and folds to produce a precise three-dimensional shape. Chemical bonds and hydrophobic interactions between R groups maintain this final tertiary structure of the protein (Figure 2.8).

A protein may be made up of several polypeptide chains held together. For example haemoglobin, the protein found in red blood cells that carries oxygen, is made up of four polypeptide chains held tightly together in a structure known as the quaternary structure.

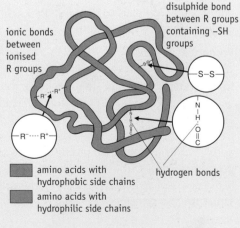

▷ **Figure 2.8** The three-dimensional structure of a protein is held in place by chemical bonds between individual amino acids and by hydrophobic interactions between R groups. Non-polar, hydrophobic side chains are arranged so they all face the inside of the molecule, excluding any water from the centre of the molecule.

Conjugated proteins

Some proteins are known as conjugated proteins – they have another chemical group associated with their polypeptide chain(s). For example, in Figure 2.9 the polypeptide chain that makes up myoglobin is associated with an iron group.

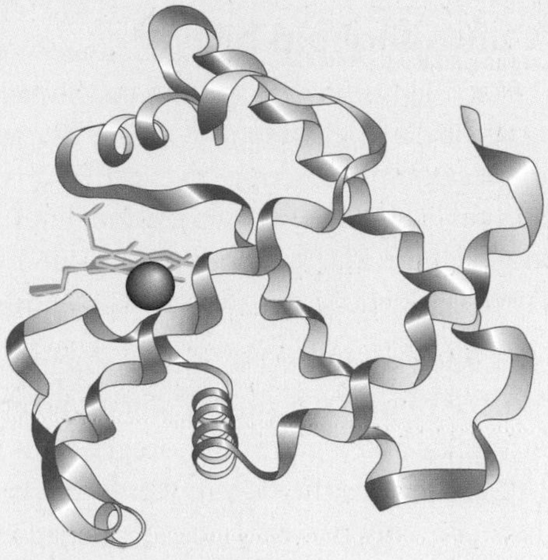

Globular and fibrous proteins

Proteins can be divided into two distinct groups:

- globular proteins
- fibrous proteins.

In **globular proteins** the polypeptide chain is folded into a compact spherical shape. These proteins are soluble due to the hydrophilic side chains that project from the outside of the molecules and are therefore important in metabolic reactions. Enzymes are an important group of globular proteins. Their three-dimensional shape is crucial to their ability to form enzyme-substrate complexes and catalyse reactions within cells.

The three-dimensional shapes of globular proteins are crucial in their function of binding to other substances, for example transport proteins within membranes and the oxygen-transport pigments haemoglobin in red blood cells and myoglobin (Figure 2.9) in muscle cells. Antibodies too are globular and rely on their precise shape to bind to the microorganisms that enter our bodies.

Figure 2.9 The globular protein myoglobin acts as an oxygen-storage molecule in muscle cells. The oxygen (red circle) attaches to the iron within the haem group (shown green). Because the protein is associated with another chemical group it is often referred to as a conjugated protein.

Fibrous proteins do not fold up into a ball shape; they remain as long chains, often with several polypeptide chains cross-linked together for additional strength. These insoluble proteins are important structural molecules. Keratin in hair and skin, and collagen (Figure 2.10) in the skin, tendons, bones, cartilage and blood vessel walls are examples of fibrous proteins.

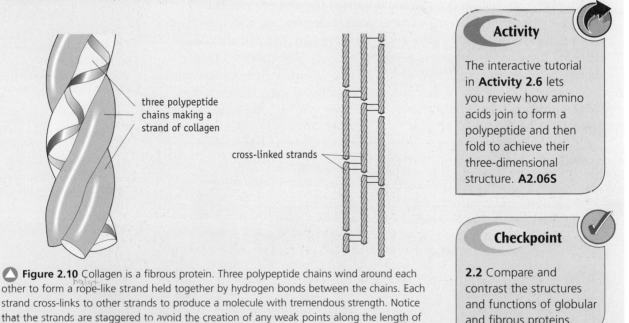

three polypeptide chains making a strand of collagen

cross-linked strands

Figure 2.10 Collagen is a fibrous protein. Three polypeptide chains wind around each other to form a rope-like strand held together by hydrogen bonds between the chains. Each strand cross-links to other strands to produce a molecule with tremendous strength. Notice that the strands are staggered to avoid the creation of any weak points along the length of the molecule.

Activity

The interactive tutorial in **Activity 2.6** lets you review how amino acids join to form a polypeptide and then fold to achieve their three-dimensional structure. **A2.06S**

Checkpoint

2.2 Compare and contrast the structures and functions of globular and fibrous proteins.

Cell membrane structure

A phospholipid bilayer

Under a light microscope the cell surface membrane looks like a single line. But closer examination using an electron microscope reveals that it is in fact a **bilayer**, appearing as two distinct lines about 7 nm wide (Figure 2.11). What do we mean by a bilayer? The basic structure is two layers of **phospholipids**.

Look back at Figure 1.42 (page 36) and remind yourself of the structure of a lipid molecule – three fatty acids and a glycerol. Compare this with Figure 2.12 and notice the difference. In a phospholipid there are only two fatty acids; a negatively charged phosphate group replaces the third fatty acid.

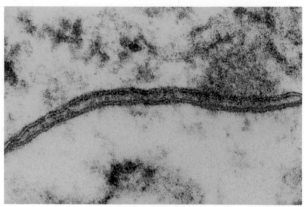

🔺 **Figure 2.11** The cell surface membrane appears as two distinct layers when viewed with an electron microscope. Can you see any membranes within the cell that have a similar structure? The same type of membrane surrounds many cell organelles.

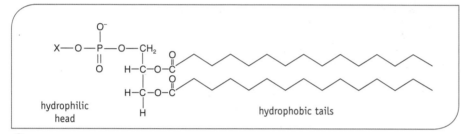

🔺 **Figure 2.12** The phospholipid molecule has two distinct sections: the hydrophilic head and the hydrophobic tails. X can be a variety of chemical groups.

The phosphate head of the molecule is **polar**, which means that the sharing of the electrons within this part of the molecule is not quite even; one end becomes slightly positive and the rest is slightly negative. This makes the phosphate head attract other polar molecules, like water, and it is therefore **hydrophilic** (water-attracting). The fatty acid tails are non-polar and therefore **hydrophobic** (water-repelling). When added to water phospholipids either form a layer on the surface with their hydrophobic tails directed out of the water, or they arrange themselves into spherical clusters called micelles to avoid contact between the hydrophobic tails and water (Figure 2.13).

Support

To find out more about polar molecules and polar bonds have a look at the Biochemistry support on the website.

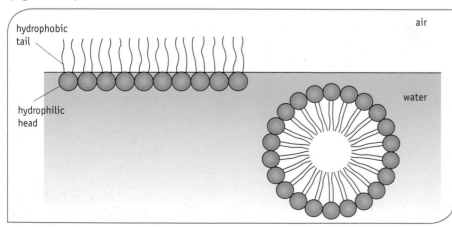

🔵 **Figure 2.13** Phospholipids in water form a monolayer on the surface or spherical micelles.

The phospholipids that make up the cell surface membrane form a bilayer with all the hydrophilic heads pointing outwards and the hydrophobic tails facing inwards avoiding any contact with water on either side of the membrane (Figure 2.14).

The fluid mosaic model

The cell surface membrane is not simply a phospholipid bilayer. It also contains proteins, cholesterol, **glycoproteins** (protein molecules with a polysaccharide attached) and **glycolipids** (lipid molecules with polysaccharides attached). Some of the proteins span the membrane. Other proteins are found only within the inner layer and some only within the outer layer (Figure 2.14). Membrane proteins have hydrophobic areas and these are positioned within the membrane bilayer.

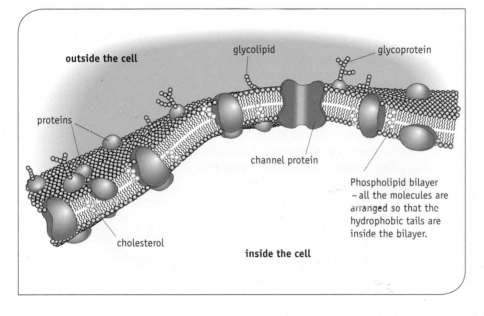

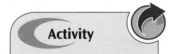 **Figure 2.14** Diagram of the fluid mosaic model of the cell surface membrane.

It is thought that some of the proteins are fixed within the membrane, but others are not and can move around in the fluid phospholipid layer. This arrangement is known as the fluid mosaic model of membrane structure. Singer and Nicholson first proposed it in 1972.

The more phospholipids containing unsaturated fatty acids there are present in the membrane, the more fluid it is. The 'kinks' in the hydrocarbon tails of the unsaturated phospholipids prevent them from packing close together so more movement is possible. Cholesterol reduces the fluidity of the membrane by preventing movement of the phospholipids.

Q2.14 Can you suggest why the membrane is more fluid with unsaturated rather than saturated phospholipids making up the bilayer?

Many different types of proteins are found within the membrane, each type having a specific function. These include functions as enzymes. As we shall see, carrier and channel proteins are involved in the transport of substances in and out of cells. Glycoproteins and glycolipids have important roles in cell-to-cell recognition and as receptors.

Activity

You can investigate the cell surface membrane practically in **Activity 2.7**. **A2.07S**

How do substances pass through cell membranes?

For a cell to function correctly it needs to be able to control transport across its surface membrane. Molecules and ions move across membranes by:

- diffusion
- osmosis
- active transport
- exocytosis
- endocytosis.

Diffusion

Diffusion is the movement of molecules or ions from an area of their high concentration to an area of their low concentration. Diffusion will continue until the substance is evenly spread throughout the whole volume. Small uncharged particles diffuse across the cell membrane, passing between the lipid molecules as they move down the concentration gradient. Small molecules like oxygen and carbon dioxide can diffuse rapidly across the cell membrane. Carbon dioxide is polar but its small size allows rapid diffusion.

Facilitated diffusion

Hydrophilic molecules and ions that are any larger than carbon dioxide cannot simply diffuse through the bilayer. They are insoluble in lipids, the hydrophobic tails of the phospholipids providing an impenetrable barrier to them. Instead they cross the membrane with the aid of proteins in a process called **facilitated diffusion**. They may diffuse through water-filled pores within **channel proteins** that span the membrane (Figure 2.14). There are different channel proteins for transporting different molecules. Each type of channel protein has a specific shape that permits the passage of only one particular molecule. Some channels can be opened or closed depending on the presence or absence of a signal which could be a specific molecule, for example a hormone, or a change in potential difference (voltage) across the membrane. These channels are called gated channels.

Some proteins that play a role in facilitated diffusion are not just simple channels but are **carrier proteins**. The ion or molecule binds onto a specific site on the protein. The protein changes shape (Figure 2.15) and as a result the ion or molecule crosses the membrane. The movement can occur in either direction, with the net movement being dependent on the concentration difference across the membrane. Molecules move from high to low concentration due to more frequent binding to carrier proteins on the side of the membrane where the concentration is higher.

Diffusion, whether facilitated or not, is sometimes called **passive transport**. 'Passive' here refers to the fact that no energy is expended in the movement.

Osmosis

Osmosis is the movement of water from a solution with a lower concentration of solute to a solution with a higher concentration of solute through a partially permeable membrane.

Activity

Activity 2.8 lets you investigate different methods of transport. **A2.08S**

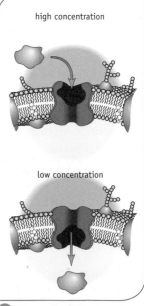

high concentration

low concentration

▲ **Figure 2.15** The carrier protein changes shape, facilitating diffusion.

Osmosis can be summarised as follows:

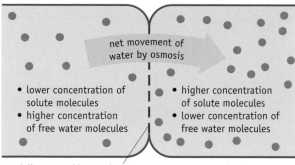

partially permeable membrane

Look at the left and right parts of Figure 2.16.
In each case decide the direction in which the solvent (i.e. water) molecules will have a net (overall) tendency to move. Only the right half of Figure 2.16 has a partially permeable membrane so only here can osmosis take place. Osmosis is due to the random movement of water molecules across the membrane and is a particular type of diffusion. If solute molecules are present, water molecules form hydrogen bonds with these solute molecules, and this reduces the movement of these water molecules. The more solute present the more water is associated with the solute and less water is free to collide with and move across the membrane.

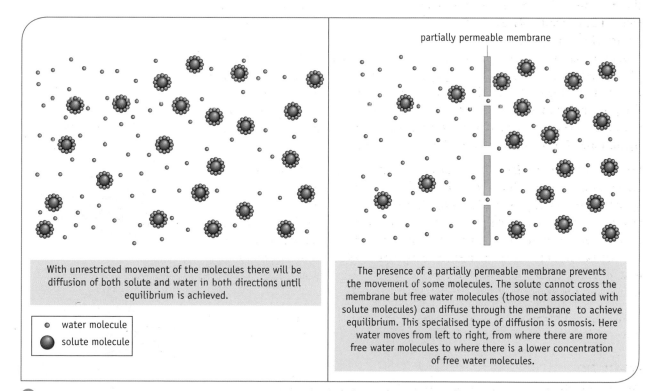

With unrestricted movement of the molecules there will be diffusion of both solute and water in both directions until equilibrium is achieved.

• water molecule
• solute molecule

The presence of a partially permeable membrane prevents the movement of some molecules. The solute cannot cross the membrane but free water molecules (those not associated with solute molecules) can diffuse through the membrane to achieve equilibrium. This specialised type of diffusion is osmosis. Here water moves from left to right, from where there are more free water molecules to where there is a lower concentration of free water molecules.

Figure 2.16 Decide which direction the water molecules will move before reading the shaded boxes to check if you were correct.

Active transport

If substances need to be moved across a membrane against a concentration gradient (from low concentration to high concentration) then energy is required. As with facilitated diffusion, specific carrier proteins are involved. The energy comes from respiration and is supplied by the energy transfer molecule **ATP**. The substance to be transported binds to the carrier protein. The energy is used to bring about the change in shape of the carrier protein and the substance is released on the other side of the membrane (Figure 2.17).

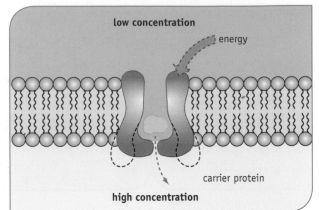

Active pumping of substances across membranes occurs in every cell. Examples appear throughout the course, including transport across epithelial cells in the lungs (later in Topic 2), plant cell roots (Topic 4), muscle cells (Topic 6), and nerve cells (Topic 8). It also occurs between compartments within a cell, for example between the mitochondria and cytoplasm (Topics 3 and 7).

Figure 2.17 In active transport, energy is required to change the shape of the carrier protein (dotted lines) and move the substance across the membrane against a concentration gradient.

Exocytosis and endocytosis

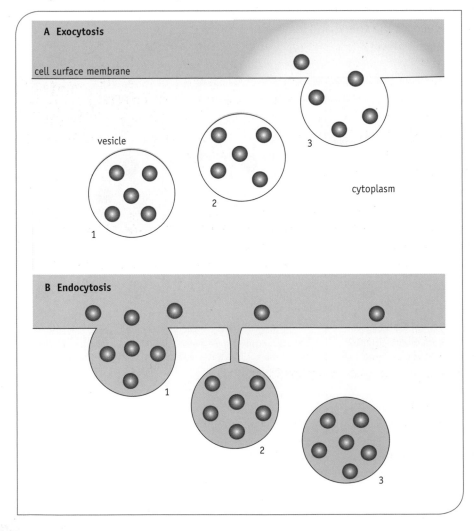

Figure 2.18 In exocytosis, vesicles fuse with the cell surface membrane, releasing their contents from the cell. In endocytosis, substances are brought inside a cell within vesicles formed from the cell surface membrane.

Bulk transport of substances across cell surface membranes is achieved by **exocytosis** and **endocytosis** (Figure 2.18). Exocytosis is the release of substances, usually proteins or polysaccharides, from the cell as vesicles fuse with the cell membrane. For example, insulin, the hormone produced by certain cells in the pancreas, is released into the blood by exocytosis. Neurotransmitter substances are also released in this way from nerve endings. Endocytosis is the reverse process: substances are taken into a cell by the creation of a vesicle. Part of the cell membrane engulfs the solid or liquid material to be transported. In some cases the substance to be absorbed attaches to a receptor in the membrane and is then absorbed by endocytosis. This, for example, is how cholesterol is taken up into cells.

Table 2.1 provides a summary of the different methods of transport across cell membranes.

▼ **Table 2.1** Summary of methods of transport across cell membranes.

Diffusion	high to low concentration until equilibrium reached hydrophobic (lipid-soluble) or small uncharged molecules through phospholipid bilayer passive, no energy required
Facilitated diffusion	high to low concentration until equilibrium reached hydrophilic molecules or ions through channel proteins or via carrier proteins that change shape passive, no energy required
Osmosis	a type of diffusion involving movement of free water molecules high to low concentration of free water molecules until equilibrium reached through phospholipid bilayer passive, no energy required
Active transport	against a concentration gradient, low to high concentration via carrier proteins that change shape requires energy
Exocytosis	used for bulk transport of substances out of the cell vesicles fuse with the cell surface membrane releasing their contents
Endocytosis	used for bulk transport of substances into the cell vesicles created from the cell surface membrane, bringing their contents into the cell

Q2.15 For each example below, suggest the type of transport most likely to be involved.

a movement of oxygen across the wall of an alveolus

b absorption of phosphate ions into root hair cells

c pumping of calcium ions into storage vesicles (small membrane-bound sacs) inside muscle cells

d release of glucose from liver cells into the bloodstream

e removal of the sodium ions that diffuse into a nerve cell, thus maintaining a low concentration within the nerve axon

f reabsorption of water molecules from the kidney tubule

What happens in the membranes of the cells lining the airways?

The cells that line the airways produce mucus. In people who do not have cystic fibrosis, the amount of water in the mucus is continuously regulated to maintain a constant viscosity ('stickiness') of the mucus. It must be runny enough to be moved by the beating cilia but not so runny that the fluid floods the airway. This regulation of the water content of the mucus is achieved by the transport of sodium ions and chloride ions across the epithelial cells. Water then follows the ions because of osmosis.

The theory below that explains this regulation is supported by evidence. Studies have been done in which the ionic composition of the airway surface liquid is measured in CF and normal mice, and in CF and healthy humans. In addition, cells from the airways have been grown in culture in the laboratory and this has given much insight into the regulation of the liquid layer on the airway surface.

Regulating water in the mucus in unaffected lungs

Excess water in the mucus

If the mucus layer contains too much water this is detected by the membranes of cells lining the airways (epithelial cells). Sodium–potassium pumps in the basal membranes of the epithelial cells actively pump sodium ions out of the cell (Figure 2.19A). The concentration of sodium ions (Na^+) in the cell falls, setting up a concentration gradient across the apical membrane (the membrane facing the airway). Sodium ions diffuse down this concentration gradient. The ions pass into the cell by facilitated diffusion through sodium channels in the apical membrane.

The raised concentration of Na^+ in the tissue fluid on the basal membrane side of the epithelial cells creates a potential difference between this tissue fluid and the mucus on the apical membrane side. The tissue fluid now contains more positively charged ions than does the mucus. This creates an electrical gradient between the tissue fluid and the mucus. This electrical gradient causes negatively charged chloride ions to diffuse out of mucus into the tissue fluid via the gaps between neighbouring epithelial cells.

The elevated Na^+ and Cl^- concentrations in the tissue fluid draw water out of the cell by osmosis across the basal membrane into the tissue fluid. This water loss increases the overall solute concentration within the cell. Since the solute concentration is now higher within the cell than in the mucus, water is drawn out of the mucus by osmosis across the apical membrane and into the cell.

Having too much water in the mucus, as described above, is a normal state of affairs. This is because the cilia are continuously moving mucus along the airways. Movement of mucus from numerous smaller bronchiole branches into fewer larger bronchioles means that water must be removed to reduce the volume of mucus so as to avoid the larger airways flooding with fluid.

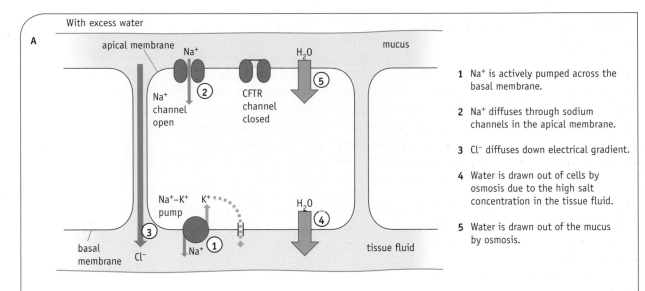

A With excess water

1 Na⁺ is actively pumped across the basal membrane.

2 Na⁺ diffuses through sodium channels in the apical membrane.

3 Cl⁻ diffuses down electrical gradient.

4 Water is drawn out of cells by osmosis due to the high salt concentration in the tissue fluid.

5 Water is drawn out of the mucus by osmosis.

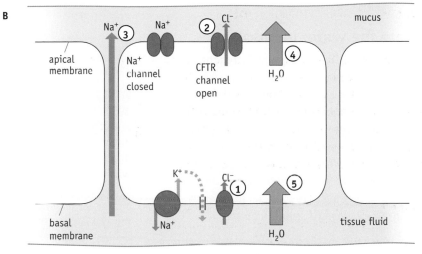

B With too little water

1 Cl⁻ is pumped into the cell across the basal membrane.

2 Cl⁻ diffuses through the open CFTR channels.

3 Na⁺ diffuses down the electrical gradient into the mucus.

4 Elevated salt concentration in the mucus draws water out of the cell by osmosis.

5 Water is drawn into the cell by osmosis.

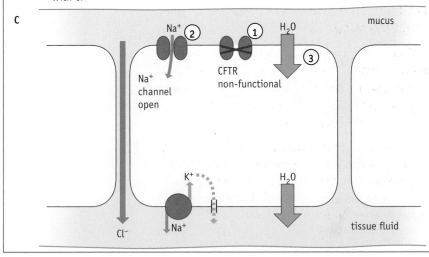

C With CF

1 CFTR channel is absent or not functional.

2 Na⁺ channel is permanently open.

3 Water is continually removed from mucus by osmosis.

Figure 2.19 The role of the CFTR channel in keeping the mucus runny.

Too little water in the mucus

When there is too little water in the mucus, chloride ions are transported across the basal membrane into the epithelial cell (Figure 2.19B). This creates a concentration gradient across the apical membrane, with the concentration of chloride ions being higher inside the cell than out. At the same time, the cystic fibrosis transmembrane regulatory (CFTR) protein channels (gated channels) open. Chloride ions diffuse out of the cell through the CFTR channels down this concentration gradient into the mucus. When open, the CFTR channels block (close) the sodium ion channels in the apical membrane. The mechanism for this is currently unknown. The build-up of negatively charged chloride ions in the mucus creates an electrical gradient between the mucus and the tissue fluid. Sodium ions diffuse out of the tissue fluid and move down this electrical gradient, passing between the cells into the mucus. The movement of the sodium and chloride ions into the mucus draws water out of the cells by osmosis until the solutions on either side of the membrane have the same concentration of free water molecules (until they are isotonic). This movement of water prevents the mucus that lines the airways from becoming too viscous (sticky).

Why CF lungs cannot regulate the water in mucus

In a person who has CF, the CFTR protein may be missing, or if present it does not function correctly (Figure 2.19C). When there is too little water in the mucus, Cl^- cannot be secreted across the apical membrane, and there is no blockage of the epithelial sodium ion channels. Since the Na^+ channels are always open, there is continual Na^+ absorption by the epithelial cells. The raised levels of Na^+ draw chloride ions and water out of the mucus into the cells. This makes the mucus more viscous, which makes it harder for the beating cilia to move it, so the mucus is not effectively cleared up and out of the lungs. Sticky mucus builds up in the airways. This mucus frequently becomes infected with bacteria, causing a downward spiral of airway inflammation and damage.

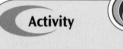

Activity

Work through the interactive tutorial in **Activity 2.9** to investigate the effects of a functioning and a non-functioning CFTR protein channel on salt and water secretion in the airways. **A2.09S**

2.3 How does CF affect other body systems?

The effect of CF on the digestive system

Cystic fibrosis sufferers have difficulty maintaining body mass because of problems with the digestion and absorption of nutrients. They also have high basal metabolic rates. They generally have poor appetites but still have to eat more than most people, including high-energy food, to make sure they obtain sufficient nutrients and energy. They require 120–140% of the recommended daily energy intake. People with CF may also take food supplements that contain digestive enzymes. The aim of these supplements is to help break down large food molecules.

Most of the chemical breakdown of food molecules and the subsequent absorption of the soluble products into the bloodstream occurs in the small intestine. Glands secrete digestive enzymes into the lumen of the gut, where they act as catalysts to speed up the breakdown. A wide range of enzymes are produced by exocrine glands (Figure 2.20) outside the gut, e.g. salivary glands, the liver and the pancreas. Enzymes are also built into the membranes of the gut wall.

Groups of pancreatic cells produce enzymes that help in the breakdown of proteins, carbohydrates and lipids. These digestive enzymes are delivered to the gut in pancreatic juice released through the pancreatic duct (Figure 2.21).

In a person with CF, the pancreatic duct becomes blocked by sticky mucus, impairing the release of digestive enzymes. The lower concentration of enzymes within the small intestine reduces the rate of digestion. Food is not fully digested, so not all the nutrients can be absorbed, and energy is lost in the faeces. This is called malabsorption syndrome.

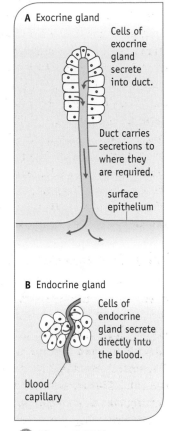

A Exocrine gland

Cells of exocrine gland secrete into duct.

Duct carries secretions to where they are required.

surface epithelium

B Endocrine gland

Cells of endocrine gland secrete directly into the blood.

blood capillary

▲ **Figure 2.20** There are two types of gland. These are: **A** exocrine glands with ducts and **B** endocrine glands which are ductless.

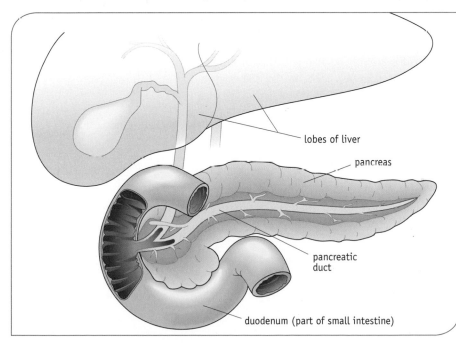

lobes of liver

pancreas

pancreatic duct

duodenum (part of small intestine)

◄ **Figure 2.21** Thick mucus can easily block the pancreatic duct in a CF patient.

An additional complication occurs when the pancreatic enzymes become trapped behind the mucus block in the pancreatic duct. These enzymes damage the pancreas itself. Cells within the pancreas produce the hormone insulin, involved in the control of blood sugar levels, and if these become damaged this can lead to a form of diabetes.

Key biological principle: Enzyme function depends on protein three-dimensional structure

Enzymes are globular proteins that act as biological **catalysts**. They speed up chemical reactions that would otherwise occur very slowly at the temperature within cells. The precise three-dimensional shape adopted by an enzyme includes a depression on the surface of the molecule called the **active site**. This site may be a relatively small part of the large protein molecule, as you can see in Figure 2.22. Only a few amino acids may be directly involved in the active site, with the remainder of the molecule maintaining the three-dimensional shape of the protein molecule.

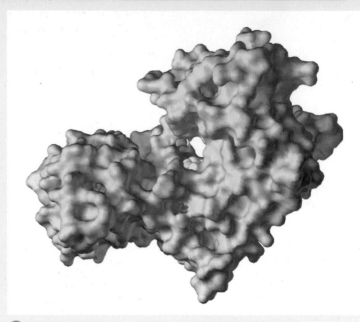

Figure 2.22 The active site is only a small part of the enzyme molecule. Here, the active site of the enzyme hexokinase changes shape as the glucose (yellow) enters.

Lock-and-key theory

Either a single molecule with a complementary shape, or more than one molecule that together have a complementary shape, can fit into the active site (Figure 2.23). These **substrate molecule(s)** form temporary bonds with the amino acids of the active site to produce an **enzyme-substrate complex**. The enzyme holds the substrate molecule(s) in such a way that they react more easily. When the reaction has taken place the products are released, leaving the enzyme unchanged. The substrate is often likened to a 'key' which fits into the enzyme's 'lock', so this is known as the **lock-and-key theory** of enzyme action. Each enzyme will only catalyse one specific reaction because of its precisely shaped active site.

Induced fit theory

It has been found that the active site is often flexible. When the substrate (or substrates) enters the active site, the enzyme molecule changes shape, fitting more closely around the substrate (Figure 2.22). It is like a person putting on a wetsuit; the wetsuit shape changes to fit the body but returns to its original shape when taken off. This is known as the **induced fit theory** of enzyme action. Only a specific substrate will induce the change in shape of an enzyme's active site.

Activation energy

To convert substrate(s) into product(s), bonds must change both within and between molecules. Breaking chemical bonds requires energy. The energy needed to break bonds and start the reaction is known as the **activation energy**. Without an enzyme, heating a substrate would provide this energy. (Think about starting a bonfire – a reaction between the chemicals in wood and oxygen. You must first provide some energy to start the fire.) The heat energy agitates atoms within the molecules; the molecules become unstable and the reaction can then proceed. In cells, enzymes reduce the amount of energy needed to bring about a reaction; this allows reactions to occur without raising the temperature of the cell.

How do enzymes reduce the activation energy?

The specific shape of the enzyme's active site and of its complementary substrate(s) is such that electrically charged groups on their surfaces interact (Figure 2.23).

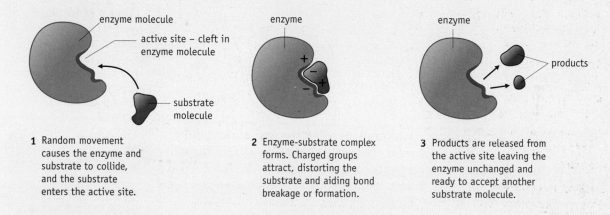

1 Random movement causes the enzyme and substrate to collide, and the substrate enters the active site.

2 Enzyme-substrate complex forms. Charged groups attract, distorting the substrate and aiding bond breakage or formation.

3 Products are released from the active site leaving the enzyme unchanged and ready to accept another substrate molecule.

Figure 2.23 Charged side groups interact and assist in breaking or forming bonds.

The attraction of oppositely charged groups may distort the shape of the substrate(s) and assist in the breaking of bonds or formation of new bonds. In some cases, the active site may contain amino acids with acidic side chains; the acidic environment created within the active site may provide conditions favourable for the reaction.

As we have seen, enzymes:

• are globular proteins
• have an active site that allows binding with a specific substrate
• catalyse (speed up) reactions
• reduce the activation energy required for a chemical reaction to take place
• do not alter the end-product or nature of a reaction
• remain unchanged at the end of a reaction, able to bind with another substrate molecule.

Finding rates of enzyme-controlled reactions

The rate of reaction is measured by determining the quantity of substrate used or the quantity of product formed in a given time. For example, when the enzyme catalase is used to break down hydrogen peroxide, H_2O_2, to water and oxygen, the rate of reaction can be found by measuring the volume of oxygen given off in a known time.

Checkpoints

2.3 Write a definition for each of the following key enzyme terms:
• biological catalyst
• activation energy
• active site
• enzyme-substrate complex
• product
• lock-and-key theory
• induced fit theory.

2.4 Explain how the three-dimensional structures of proteins enable enzymes to perform their functions as biological catalysts.

If we mix a fixed quantity of enzyme and substrate, at first the reaction will proceed quickly, as shown in Figure 2.24. However, as the substrate is used up, there are fewer substrate molecules to bind with the enzyme and the reaction slows down and eventually stops (no further increase in the product occurs), as seen in Figure 2.25B. The slope of the rapid phase of the reaction is known as the initial rate of reaction and is frequently used when comparing rates of enzyme-controlled reactions.

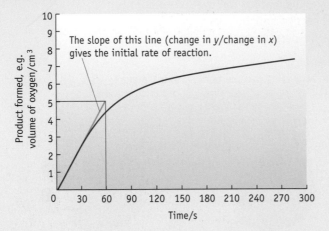

Figure 2.24 The quantity of product is measured over time to determine the progress of an enzyme-catalysed reaction. Here the volume of oxygen produced is measured when catalase is added to hydrogen peroxide.

How do enzyme and substrate concentrations affect the rate of reaction?

Figure 2.25 shows the effect of enzyme concentration on the initial rate of reaction. The initial rate of reaction is directly proportional to the enzyme concentration because the more enzyme that is present, the greater the number of active sites that are available to form enzyme-substrate complexes. The increase in rate will continue in this linear fashion assuming that there is an excess of substrate. In Figure 2.25B you can see how at high substrate concentrations it is the enzyme concentration that limits the rate of reaction. Every active site is occupied and substrate molecules cannot enter an active site until one becomes free again.

Activity

In **Activity 2.10** you can investigate the effect of enzyme concentration on enzyme activity. **A2.10S**

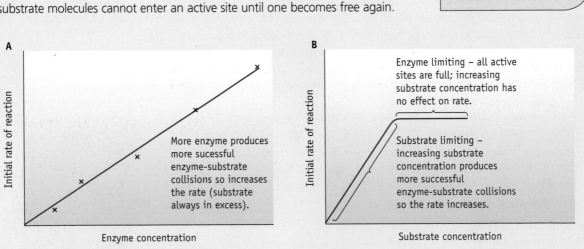

Figure 2.25 The effect of **A** enzyme concentration and **B** substrate concentration on the rate of an enzyme-controlled reaction.

The effect of CF on the reproductive system

We have seen how sticky mucus produced by a defective CFTR protein can lead to major complications in the lungs and pancreas. In the reproductive system it can also cause severe problems. Females have a reduced chance of becoming pregnant because a mucus plug develops in the cervix (Figure 2.26A). This stops sperm from reaching the egg. Males with cystic fibrosis commonly lack the vas deferens (sperm duct) on both sides, which means sperm cannot leave the testes (Figure 2.26B). Where the vas deferens is present it can become partially blocked by a thick sticky mucus layer. This means fewer sperm are present in each ejaculate.

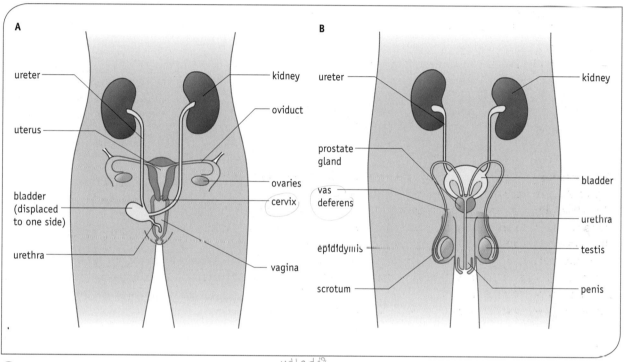

Figure 2.26 A The female reproductive system. In CF a mucus plug develops in the cervix. **B** The male reproductive system. In CF the vas deferens may be absent or plugged with mucus.

Q2.16 Salty sweat is often one of the first signs that a baby may have CF. Why might the sweat of a person with CF be more salty than normal? (Hint: in sweat ducts the CFTR protein works in the opposite direction to the way it works in mucus-producing cells.)

Checkpoint

2.5 For each of a to c, write five bullet point statements to describe the effect CF has on:
a the gas exchange system **b** the digestive system **c** the reproductive system.

2.4 How is the CFTR protein made?

Cystic fibrosis is caused by an error in the DNA that carries the instructions for making the CFTR protein. In order to know how DNA works (and what has gone wrong in cystic fibrosis), you need to know what a gene is and how the codes it contains are used to make proteins.

Did you know? DNA and immortality

'The seed of life itself. Peel the chains apart, each chain reproduces the other, one becomes two, two become one. Generation on generation, all the way from Adam and Eve to you and me. It never dies. One simple shape. The womb of humanity. Endlessly, effortlessly fertile, dividing, reforming ... It's the closest we'll ever get to immortality.'

(*Source:* Tim Piggott-Smith, who played Francis Crick in the BBC film *Life Story*, talking about the structure of DNA.)

In 1953 James Watson and Francis Crick (Figure 2.27) proposed a model for the structure of DNA, using the X-ray diffraction patterns of Rosalind Franklin and Maurice Wilkins. Their model was correct and their discovery has revolutionised biology.

DNA is found in every cell nucleus. It contains the genetic code which dictates all our metabolism and inherited characteristics. It does this by controlling the manufacture of proteins. Your proteins are what

🔺 **Figure 2.27** James Watson (left) and Francis Crick (right) worked out the structure of DNA in 1953.

make you unique; they are what make you a human being and not an oak tree or a chimpanzee. They play a vital role in giving you the unique characteristics that mean you are not the same as the person sitting next to you (you may be rather glad about that!).

The structure of DNA

Gene and genome

A **gene** is a segment of DNA which codes for one protein. Each chromosome found in the cell nucleus contains DNA and carries numerous genes. The genes make up only a fraction of the total length of DNA in the chromosomes; the job of the remainder of the DNA is not fully known. Together, all the genes in an individual (or species) are known as the **genome**.

Extension

You can read about the controversy surrounding the work of Rosalind Franklin and DNA in **Extension 2.1**. **X2.01S**

DNA is a chain of nucleotides

DNA is one type of nucleic acid, called **deoxyribonucleic acid**. It is a long chain molecule made of many units called nucleotides.

Q2.17 DNA is called a polynucleotide. Explain why.

A nucleotide contains three molecules linked together by condensation reactions: **deoxyribose** (a 5-carbon sugar), a **phosphate group** and an

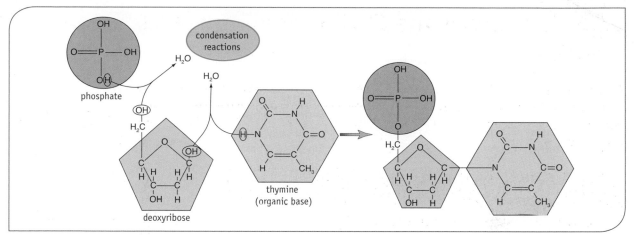

🔺 **Figure 2.28** A phosphate, a deoxyribose sugar and a base join to form a nucleotide.

organic base containing nitrogen. Look at Figure 2.28 to see how these three are arranged in a nucleotide.

Nucleotides link together by condensation reactions between the sugar of one nucleotide and the phosphate of the next one, producing a long chain of nucleotides. The nitrogen-containing base is the only part of the nucleotide that is variable. There are four bases – **adenine**, **cytosine**, **guanine** and **thymine**. These bases are often represented by their initial letter: **A**, **C**, **G** and **T** respectively.

In a DNA molecule there are two long strands of nucleotides twisted around each other to form a double helix (Figure 2.29), rather like a spiral staircase. The sugars and phosphates form the 'backbone' of the molecule and are on the outside. The bases point inwards horizontally, rather like the rungs of a ladder. The two strands which run in opposite directions are known as antiparallel strands and are held together by hydrogen bonds between pairs of bases. The DNA in each human cell contains some 3000 million of these base pairs.

Why do the bases pair up?

If you look at Figure 2.29 you should notice that the bases only pair in a certain way: adenine only pairs with thymine, and cytosine only pairs with guanine.

Q2.18 Look at Figure 2.29 and suggest why the bases might only form these pairs.

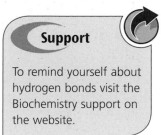

Support

To remind yourself about hydrogen bonds visit the Biochemistry support on the website.

🔻 **Figure 2.29** The structure of DNA. Remember that any picture that you see of DNA is a simplification, a model based on evidence from techniques such as X-ray diffraction. A whole DNA molecule is much longer than this and usually contains millions of base pairs.

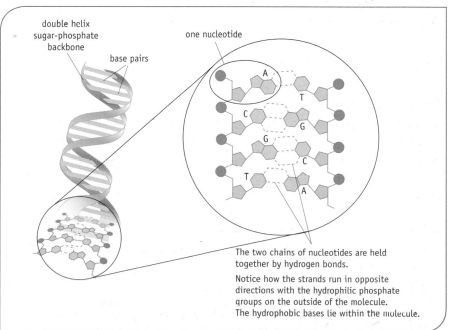

The two chains of nucleotides are held together by hydrogen bonds.

Notice how the strands run in opposite directions with the hydrophilic phosphate groups on the outside of the molecule. The hydrophobic bases lie within the molecule.

The key to this pairing is in the structure of the bases and the bonding between them. Bases A and G both have a two-ring structure, whereas C and T have only one ring. The bases pair so that there are effectively three rings forming each rung of the DNA molecule, making the molecule a uniform width along its whole length. The shape and chemical structure of the bases dictates how many bonds each can form and this determines the pairing of A with T (two hydrogen bonds) and C with G (three hydrogen bonds). This deceptively simple fact is the clue to how DNA works. The bases A and T and the bases C and G are referred to as complementary base pairs.

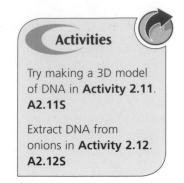

Q2.19 The sequence of bases on part of one strand of a DNA molecule is:

A T C C C T G A G G T C A G T

What would be the sequence of bases on the corresponding part of the other strand?

How does DNA code for proteins?

The CF gene is on chromosome 7. It is a long gene, made up of about 230 kbp (230 000 base pairs), and it instructs the cell to make the CFTR protein that forms the transmembrane chloride channel. But what is the genetic code? The sequence of bases in the DNA tells the cell which amino acids to link together to make the CFTR protein.

The triplet code

In the genetic code, one base does not simply code for one amino acid. There are only four bases, so if this were the case proteins could contain only four different amino acids, instead of the 20 amino acids found commonly in proteins. The code carried by the DNA is a three-base or **triplet code** (Figure 2.30). Each adjacent group of three bases codes for an amino acid; the triplets do not overlap. There are 64 three-letter combinations possible; some are start signals and others stop signals (called chain terminators). Several triplets code for the same amino acid; the code is therefore known as degenerate. In some cases all the codes with the same first two letters code for the same amino acid. This amazingly simple but fundamental coding system is found in all organisms.

From DNA to proteins

The DNA in our chromosomes carries the genetic information from one generation to the next. It carries the codes that help determine the structure and function of the cell by telling the cell which proteins to make. But DNA is in the nucleus, and proteins are made in the cytoplasm. DNA cannot pass through the membranes surrounding the nucleus into the cytoplasm. So how do the instructions get from nucleus to cytoplasm?

A 'copy' of the DNA is made. It is rather like making a photocopy of some precious original plans so that everyone on the factory floor can work from them, leaving the originals safe at head office. This 'copy' is not DNA but another type of nucleic acid called **ribonucleic acid**, **RNA**. This RNA leaves the nucleus, carrying the information to the cytoplasm where it is used in the manufacture of proteins.

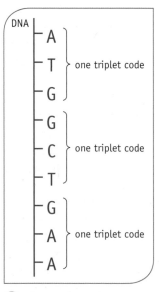

▲ **Figure 2.30** Each DNA triplet of three bases codes for one amino acid. The triplet codes do not overlap.

What are the differences between DNA and RNA?

An RNA molecule has a *single* strand made of a string of RNA nucleotides. These are very similar in structure to DNA nucleotides except that they contain **ribose** sugar and not deoxyribose (Figure 2.31). Another difference is that in RNA nucleotides, the base **uracil** (**U**) replaces thymine, so RNA never contains thymine.

There are three types of RNA involved in protein synthesis: **messenger RNA** (**mRNA**), **transfer RNA** (**tRNA**) and **ribosomal RNA** (**rRNA**). Their different roles in protein synthesis can be seen in Figures 2.35 and 2.36 on pages 78–80.

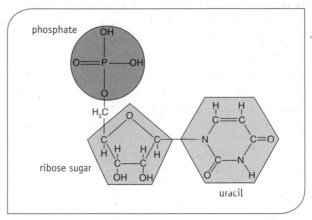

▲ **Figure 2.31** Compare this RNA nucleotide with the DNA nucleotide in Figure 2.28 and notice the differences.

Protein synthesis

Protein synthesis has two stages; the first occurs in the nucleus and the second in the cytoplasm (Figure 2.32).

1 Transcription

Transcription takes place in the nucleus. The DNA double helix unwinds and hydrogen bonds between the bases break, allowing the two strands to partly separate. The sequence on one of the strands, the **template strand**, is used in the production of a messenger RNA molecule. This mRNA is built from free RNA nucleotides which line up alongside the DNA strand. Because of base pairing the order of bases on the DNA exactly determines the order of the bases on the RNA. In other words, every triplet code on DNA gives rise to a complementary **codon** on messenger RNA. This complementary base pairing can be seen in Figure 2.33. This process of synthesising mRNA is shown in Figure 2.35 and involves a number of enzymes such as RNA polymerase. The completed messenger RNA molecule now leaves the nucleus through a pore in the **nuclear envelope** (the two membranes that surround the nucleus) and enters the cytoplasm. This is where the second stage takes place.

2 Translation

Translation starts once the messenger RNA molecule becomes attached to a **ribosome**. Ribosomes are small organelles (structures within the cell) made of ribosomal RNA and protein. Ribosomes are found free in the cytoplasm or attached to **endoplasmic reticulum**, a system of flattened, membrane-bound sacs. A transfer RNA molecule carrying an amino acid molecule also becomes attached to the ribosome. The tRNA molecule has three bases called an anticodon (Figure 2.33) and these pair with complementary bases on the mRNA codon. Then the amino acid that the tRNA has carried becomes attached to a growing chain of amino acids by means of a **peptide bond**. The ribosome moves along the messenger RNA until all the codons have been used and the complete chain of amino acids has been produced. See Figure 2.36. (The mRNA codons for each amino acid are shown in Figure 2.41, page 83.)

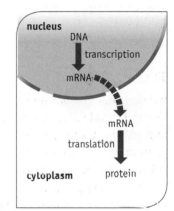

▲ **Figure 2.32** In transcription, the DNA code is copied onto mRNA. In translation, it is used to join the correct sequence of amino acids and build a new protein.

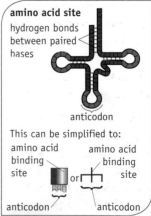

▶ **Figure 2.33** The tRNA molecule carries an amino acid. Its anticodon, base pairs with the codon on the mRNA molecule.

Transcription and translation – a little more detail

Figure 2.34 outlines the relationship between the base codes on DNA, mRNA and tRNA. In transcription, an enzyme called **RNA polymerase** attaches to the DNA (Figure 2.35). The hydrogen bonds between paired bases break and the DNA molecule unwinds. RNA nucleotides with bases complementary to those on the template strand of the DNA pair up and bond to form an mRNA molecule. When complete, the mRNA molecule leaves the nucleus through a pore in the nuclear envelope.

The template strand is also known as the antisense strand because once transcribed it makes an mRNA molecule with the same base sequence as the DNA coding strand. The coding strand is known as the sense strand. Be aware that many biology text books are incorrect in their use of the terms sense and antisense strand.

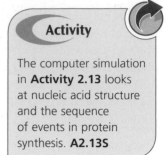

Activity

The computer simulation in **Activity 2.13** looks at nucleic acid structure and the sequence of events in protein synthesis. **A2.13S**

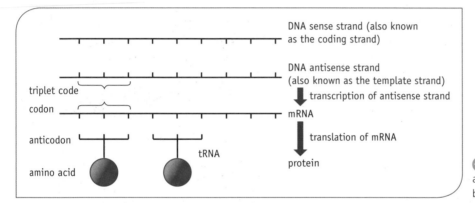

Figure 2.34 Codons, anticodons and complementary base pairing.

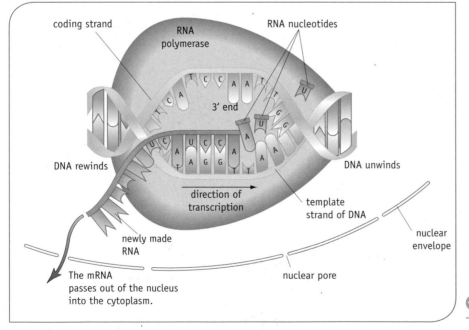

Figure 2.35 Protein synthesis – transcription.

- The mRNA attaches to the surface of a ribosome. Ribosomes are made up of two subunits. The mRNA attaches to the smaller subunit so that two mRNA codons face the two binding sites of the larger subunit.

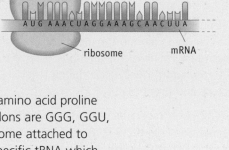

At one side of the tRNA is a triplet base sequence called an anticodon. The three bases of the anticodon are complementary to the mRNA codon for an amino acid. For example, the mRNA codons for the amino acid proline are CCC, CCA, CCG and CCU. The complementary anticodons are GGG, GGU, GGC and GGA. Within the cytoplasm free amino acids become attached to the correct tRNA molecules. Each amino acid has its own specific tRNA which carries it to the ribosome.

- The first codon exposed on the ribosome is always the start code AUG. This codes for the amino acid methionine. The tRNA molecule with the complementary anticodon UAC hydrogen bonds to the codon. The next codon is facing the binding site. This codon attracts the tRNA–amino acid complex that has the complementary anticodon and it binds.

- The ribosome holds the mRNA, tRNAs, amino acids and associated enzyme in place while a peptide bond forms between the two amino acids. The peptide bond is a condensation reaction between the amino group of one amino acid and the carboxylic acid group of the next, forming a dipeptide.

- Once the peptide bond has formed the ribosome moves along the mRNA to reveal a new codon at the binding site. The first tRNA returns to the cytoplasm.

- The whole process is repeated and translation continues until the ribosome reaches a stop signal: UAA, UAC or UGA.

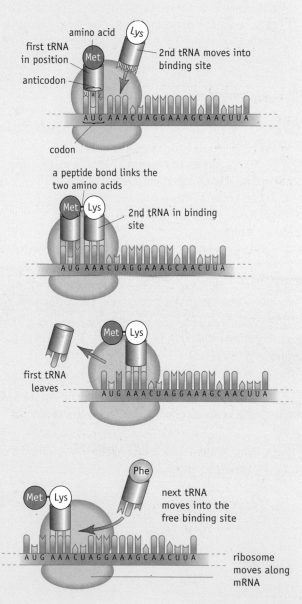

Figure 2.36 Protein synthesis – translation.

A protein molecule can be made up of a combination of thousands of the 20 naturally occurring amino acids. The sequence of the amino acids, its primary structure, determines the structure and properties of the protein and a slight variation in the order, for example changing even one of the amino acids in the chain, may substantially alter the protein's structure and properties.

If several ribosomes attach to a single mRNA molecule (see Figure 2.37), several copies of the same protein can be produced at the same time. As the protein molecule is formed, it folds up into the three-dimensional shape determined by its primary structure – the sequence of amino acids. The amino acids in different parts of the chain interact with each other forming bonds which hold the protein molecule in a precise shape.

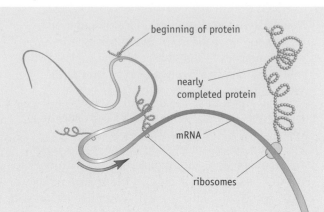

🔺 **Figure 2.36** continued

Q2.20 On which type of RNA would you find **a** a codon **b** an anticodon?

Q2.21 The sequence of bases AGT form a triplet code on the sense strand. What is:
a its triplet code on the antisense strand
b its codon
c its anticodon
d the amino acid it codes for? (Use the table of mRNA codons in Figure 2.41 on page 83 to answer this.)

Q2.22 Using the table of mRNA codons in Figure 2.41 work out the sense strand triplet code for tryptophan.

Q2.23 What would be the sequence of bases on a length of messenger RNA built using the following DNA strand as a template?

T A C A T G G A T T C C G A T

Q2.24 How many tRNA molecules would be involved in the synthesis of the protein coded for by this section of DNA?

Q2.25 What are the anticodons, assuming you read the section from left to right?

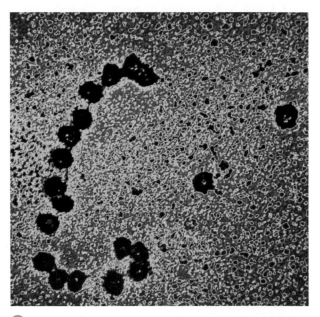

🔺 **Figure 2.37** False-colour transmission electron micrograph showing several ribosomes on an mRNA strand.

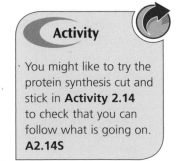

Activity

You might like to try the protein synthesis cut and stick in **Activity 2.14** to check that you can follow what is going on. **A2.14S**

2.5 What goes wrong with DNA?

A mistake in transcription can produce mRNA with one or more incorrect codons. This could result in the production of a faulty protein or no protein at all. But because the fault is in the mRNA, it would only affect the proteins produced from this one mRNA strand in this one cell, on this one occasion. It would not produce the problems seen in every epithelial cell of cystic fibrosis sufferers. It is errors in the DNA that are responsible for inherited genetic conditions. These mistakes arise when DNA copies itself during the process of DNA replication.

DNA replication

When a cell divides, an exact copy of the DNA must be produced so that each of the daughter cells receives a copy. This process of copying the DNA is called **replication**.

The DNA double helix unwinds from one end and the two strands split apart as the hydrogen bonds between bases break (Figure 2.38). Free DNA nucleotides line up alongside each DNA strand and hydrogen bonds form between the complementary bases. The enzyme DNA polymerase links the adjacent nucleotides to form a complementary strand. In this way each strand of DNA acts as a template on which a new strand is built and, overall, two complete DNA molecules are formed. These are identical to each other and to the original DNA molecule. Check this by comparing them in Figure 2.38. Each of the two DNA molecules now contains one 'old' strand and one 'new' strand. This process is therefore known as **semi-conservative replication**.

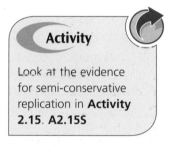

Activity

Look at the evidence for semi-conservative replication in **Activity 2.15**. **A2.15S**

▼ **Figure 2.38** DNA replication.

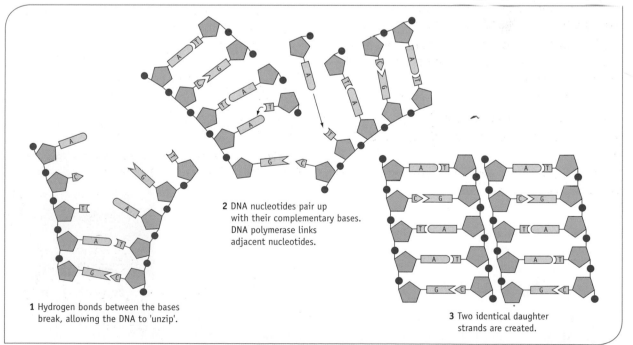

1 Hydrogen bonds between the bases break, allowing the DNA to 'unzip'.

2 DNA nucleotides pair up with their complementary bases. DNA polymerase links adjacent nucleotides.

3 Two identical daughter strands are created.

Mistakes in replication

So what has all this to do with cystic fibrosis and other genetic diseases? Sometimes DNA replication does not work perfectly. As the 'new' strand of DNA is being built, an incorrect base may slip into place. This is an example of a gene **mutation**.

Q2.26 Assuming no mutations, what would be the sequence of bases on the complementary strand created by replication of an 'old' strand with the sequence C A G T C A G G C?

Q2.27 Identify the mutations that have occurred in replication of the same sequence in each of the following:

a G T C A G G C C G   **d** G T C G T C C G
b G A C A G T C C G   **e** G T C A G G T C C G
c G T C A T G C C G

Sometimes mutations occur in the DNA of an ovary or testis cell that is dividing to form an egg or sperm. Such a mutation may be passed on to future generations, present in every single cell produced from the fertilised egg.

Some mutations have no effect on the organism. Large amounts of the DNA found in the cell does not actually play a role in protein synthesis and therefore mutations that occur in these sections may have no effect. However, if a mutation occurs within a gene and a new base triplet is created that codes for a stop signal or a different amino acid, the protein formed may be faulty. This could cause a genetic disorder.

A mutation causes sickle cell anaemia

In the disease **sickle cell anaemia**, there is a mutation in the gene that codes for one of the polypeptide chains in haemoglobin, the pigment in red blood cells which carries oxygen around the body. The base adenine replaces thymine at one position along the chain. The mRNA produced from this DNA contains the triplet code GUA rather than GAA. As a result the protein produced contains the amino acid valine rather than glutamic acid at this point. This small change has a devastating effect on the functioning of the molecule. The haemoglobin is less soluble. When oxygen levels are low, the molecule forms long fibres that stick together inside the red blood cell, distorting its shape. The half moon (sickle) shaped cells carry less oxygen and can block blood vessels (Figure 2.39).

Mutations and cystic fibrosis

The CF gene is a section of DNA carrying the code to make the CFTR protein. It is located on chromosome 7. The protein it codes for is 1480 amino acids long and these are arranged into the 3D structure shown in Figure 2.40.

Cystic fibrosis is not as straightforward a genetic story as sickle cell anaemia. In the CF gene hundreds of different mutations have been identified that can give rise to cystic fibrosis. The mutations affect the CFTR protein in different ways. In some cases ATP is unable to bind and open the ion channel; in other cases the channel is open but changes in the protein structure lead to reduced movement of chloride ions through the channel. The mutations are passed from parent to offspring. The most common mutation, known as the DF508 mutation, is the deletion of three nucleotides (Figure 2.41). This

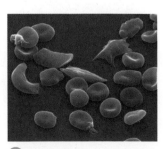

Figure 2.39 Can you identify the sickle-shaped red blood cells, symptoms of sickle cell anaemia?
Magnification ×1650.

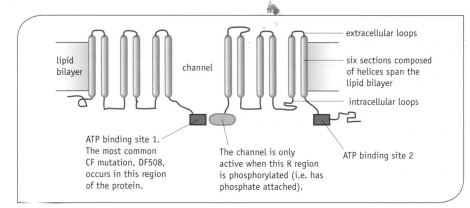

Figure 2.40 The CFTR protein is a channel protein.

Weblink

You can find out more about the mutations that cause cystic fibrosis by visiting the Oxford Gene Medicine Research Group website.

causes the loss of phenylalanine, the 508th amino acid in the CFTR protein, which is thought to result in misfolding of the protein.

Q2.28 Look at Figure 2.41 and explain why it is only phenylalanine that is lost even though two triplet codes have lost nucleotides.

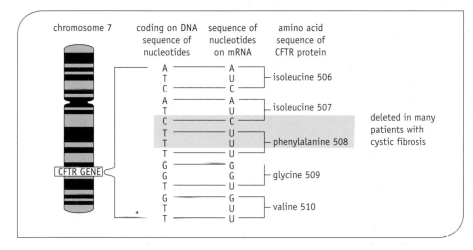

Figure 2.41 The most common mutation causing cystic fibrosis is the deletion of three bases within the CF gene. You can see the consequences of this by consulting the genetic dictionary in the table below.

mRNA codons		Second base							Third base
		U		**C**		**A**		**G**	
U	UUU	Phe	UCU	Ser	UAU	Tyr	UGU	Cys	U
	UUC	Phe	UCC	Ser	UAC	Tyr	UGC	Cys	C
	UUA	Leu	UCA	Ser	UAA	Stop	UGA	Stop	A
	UUG	Leu	UCG	Ser	UAG	Stop	UGG	Trp	G
C	CUU	Leu	CCU	Pro	CAU	His	CGU	Arg	U
	CUC	Leu	CCC	Pro	CAC	His	CGC	Arg	C
	CUA	Leu	CCA	Pro	CAA	Gln	CGA	Arg	A
	CUG	Leu	CCG	Pro	CAG	Gln	CGG	Arg	G
A	AUU	Ile	ACU	Thr	AAU	Asn	AGU	Ser	U
	AUC	Ile	ACC	Thr	AAC	Asn	AGC	Ser	C
	AUA	Ile	ACA	Thr	AAA	Lys	AGA	Arg	A
	AUG	Met	ACG	Thr	AAG	Lys	AGG	Arg	G
G	GUU	Val	GCU	Ala	GAU	Asp	GGU	Gly	U
	GUC	Val	GCC	Ala	GAC	Asp	GGC	Gly	C
	GUA	Val	GCA	Ala	GAA	Glu	GGA	Gly	A
	GUG	Val	GCG	Ala	GAG	Glu	GGG	Gly	G

Key:
Ala = alanine
Arg = arginine
Asn = asparagine
Asp = aspartic acid
Cys = cysteine
Gln = glutamine
Glu = glutamic acid
Gly = glycine
His = histidine
Ile = isoleucine
Leu = leucine
Lys = lysine
Met = methionine
Phe = phenylalanine
Pro = proline
Ser = serine
Thr = threonine
Trp = tryptophan
Tyr = tyrosine
Val = valine

Do Claire and Nathan have one of these mutations? Will they pass it on and if they do, will the child have the disease?

2.6 How is CF inherited?

Claire and Nathan know that cystic fibrosis is inherited but they need to understand *how* it is inherited. This will tell them the chances of their children inheriting the disease.

How genes are passed on

Genes and chromosome pairs

A gene is a length of DNA that codes for a protein. Every cell (except the sex cells) contains *two* copies of each gene, one from each parent. For any particular gene the two copies are located in the same position, or **locus**, one on each of two paired chromosomes. We have 23 pairs of chromosomes found in our cells, called **homologous chromosomes**. Within each of these pairs, one chromosome comes from our mother and the other from our father.

Cystic fibrosis is caused by a gene mutation passed on from parents to their children. But consider the following three situations:

1   A couple has six children. The first five are healthy; the sixth has cystic fibrosis. Neither parent has the disease.

2   Another couple has two children, both of whom have cystic fibrosis. Again, neither parent has the disease.

3   A woman with cystic fibrosis is told that it is unlikely that her children will have the disease, but they will all be 'carriers' and could themselves have children with the illness.

So what is going on?

Genotypes, phenotypes and alleles

As we have seen, cystic fibrosis is caused by a mutation in the CF gene, the length of DNA that codes for the CFTR protein. The CF gene occurs in two alternative forms or **alleles**. First, there is the normal allele which codes for the production of functioning CFTR protein; this can be represented by the letter **F**. Secondly, there is the mutated allele which produces a non-functional protein; this can be represented by the letter **f**. Since every human being has two copies of this gene in all their body cells there are three possible combinations that can occur, namely:

1   **FF** – This person has two identical copies of the normal allele and does not have cystic fibrosis.

2   **ff** – This person has two copies of the mutated allele and does have cystic fibrosis.

3   **Ff** – This person has one normal allele and one mutated allele. He or she does not have cystic fibrosis but is a **carrier**. He or she could have children who have the disease.

The alleles that a person has make up their **genotype**. Combinations 1 and 2, **FF** and **ff**, show a **homozygous** genotype for the CF gene – there are two identical copies of the allele. Combination 3 has two different alleles, **Ff**, and is **heterozygous** for the CF gene.

The characteristic caused by the genotype, i.e. its observable effect, is the **phenotype**. Table 2.2 summarises the cystic fibrosis genotypes and phenotypes.

F is called the **dominant allele**. It affects the phenotypes of both the homozygote (a person who has the homozygous genotype, **FF**) and the heterozygote (a person with the heterozygous genotype, **Ff**). On the other hand, if **f** is the **recessive allele**; it only affects the phenotype of the homozygote (**ff**).

If two CF carriers had children what genotypes and phenotypes would we expect their children to have?

Table 2.2 The relationship between genotype and phenotype at the cystic fibrosis gene.

Genotype	Phenotype
FF	normal
ff	cystic fibrosis
Ff	normal, but carrier

Predicting the genotypes of offspring

In the UK about 1 person in 24 is a cystic fibrosis carrier (**Ff**). They can pass the disease on to their children. When gametes are produced, each egg or sperm contains only *one* allele, in this case either **F** or **f**. These two types of gametes are produced in roughly equal numbers. The expected genotypes of children produced by two cystic fibrosis carriers can be shown in a genetic diagram, known as a Punnett square. This illustrates all the possible ways in which the two types of allele can combine, and thus shows the possible genotypes that can occur in the children.

| Parent's phenotypes | normal | normal |
| Parent's genotypes | **Ff** | **Ff** |

Gametes from mother

Gametes from father		F	f
	F	FF	Ff
	f	Ff	ff

Gamete genotypes (F) or (f)   (F) or (f)

This means that every time a child is born to parents who are both carriers of cystic fibrosis, there is a 1 in 4 or 0.25 probability that the baby will have the genotype **ff** and suffer from cystic fibrosis. There is also a 1 in 4 probability that the genotype will be **FF**, and a 1 in 2 (2 in 4) or 0.5 probability that it will be **Ff**, a CF carrier.

Until very recently people with the genotype **ff** did not often survive to be adults and thus did not have any children. You might think that natural selection would have eliminated such a harmful gene. In fact, heterozygotes (**Ff**) have some protection against the dangerous disease typhoid. So in areas where typhoid was common, carriers would have been at a definite advantage.

Q2.29 Give a genetic explanation for the three situations described at the top of page 84.

Some human genetic diseases

Cystic fibrosis is an example of **monohybrid inheritance**, so called because the characteristic is controlled by only one gene. Most human characteristics are inherited in a much more complex way, and are often influenced by environmental factors. However, there are a few characteristics controlled by single genes.

For example, **thalassaemia** is a genetic disease caused by recessive alleles of a gene on chromosome 11. The gene is involved in the manufacture of the protein haemoglobin, found in red blood cells, which carries oxygen around the body. A number of different mutations can affect this gene. Someone who is homozygous for one of these conditions either makes no haemoglobin at all, or makes haemoglobin that cannot carry out its function. The homozygous condition is often eventually lethal.

People who are heterozygous show no symptoms, but have some protection against malaria. For this reason thalassaemia is relatively common in people who live (or have ancestors who lived) in areas where malaria occurs or occurred in the past, particularly around the Mediterranean Sea. Because of this 'heterozygous advantage' the mutant alleles have not disappeared.

Other conditions such as **albinism**, **phenylketonuria** and **sickle cell anaemia** are also caused by single recessive alleles. **Achondroplasia**, on the other hand, is caused by a dominant allele. A homozygote, carrying two copies of the allele for achondroplasia, always dies. Someone who is heterozygous for this condition will show very restricted growth, usually attaining a height of about 125–130 cm. The trunk is of average length but the limbs are much shorter. Their mental powers are not affected. **Huntington's disease** and the ability to taste PTC (a bitter-tasting chemical called phenylthiocarbamide) are also caused by dominant alleles.

You should not suppose that characteristics determined by a single gene are found only in humans. They occur in all organisms. Indeed, they were first discovered in the garden pea by Gregor Mendel in the 1850s and 1860s.

The work of Mendel

Mendel is rightly known as the father of genetics. He was a monk and carried out a huge number of breeding experiments in the garden of his Moravian monastery. Mendel died without other scientists appreciating the significance of his work. However, he established that a number of characteristics of the garden pea were determined by separate genes. For example, whether the plants are tall (1.9–2.2 m in height) or short (0.3–0.5 m) is determined by one gene. Whether the seeds are smooth or wrinkled is determined by another gene.

Activities

You can use Reebops to investigate inheritance in **Activity 2.16**. **A2.16S**

Activity 2.17 lets you apply ideas about inheritance to some other situations. **A2.17S**

Checkpoint

2.6 Produce a vocabulary list giving definitions of the following key genetic terms:
- locus
- homologous chromosomes
- alleles
- dominant allele
- recessive allele
- heterozygous
- homozygous
- carrier
- genotype
- phenotype

2.7 How is CF treated?

The CF treatments currently available alleviate some of the symptoms. This can contribute to a longer life and improved quality of life, but as yet there is no known cure for cystic fibrosis.

Did you know? Treatments for CF

1 Medication
There are a wide range of medications commonly used to relieve the symptoms of CF. These include the following:
- Bronchodilators – these are drugs that are inhaled using a nebuliser. Air is blown through a solution to make a fine mist which is breathed into the lungs through a mouthpiece. The drugs relax the muscles in the airways, opening them up and relieving tightness of the chest.
- Antibiotics – the early diagnosis and treatment of lung infection is the cornerstone of CF treatment. Many antibiotics are used to kill or prevent the growth of bacteria in the lungs.
- DNAase enzymes – infection of the lungs leads to the accumulation of white blood cells in the mucus. The breakdown of these white blood cells releases DNA, which can add to the 'stickiness' of the mucus. DNAase enzymes can be inhaled using a nebuliser. They break down the DNA, so the mucus is easier to clear from the lungs.
- Steroids are used to reduce inflammation of the lungs.

2 Diet
Adults with CF are recommended to eat high-energy foods and their diet should include double the quantity of protein recommended for people who do not have CF. Some people with CF may also need salt supplements.

3 Digestive enzyme supplements
If the pancreatic duct is blocked, the food molecules in the small intestine cannot be broken down far enough to be absorbed. Taking enzyme supplements with food helps to complete the process of digestion.

4 Physiotherapy
Rhythmical tapping of the walls of the chest cavity (percussion therapy) can help loosen the mucus and improve the flow of air into and out of the lungs (Figure 2.42). Such treatment needs to be carried out regularly, twice a day.

5 Heart-and-lung transplant
If the lungs become badly damaged and very inefficient, other treatments may become ineffective at relieving the symptoms. The only option available may be to replace the damaged lungs with a heart-and-lung transplant.

Figure 2.42 Children can learn how to perform their own physiotherapy. Here a child uses a flutter device. Exhaling through the flutter's special valve causes rapid changes in air pressure within the airways. The vibrations aid movement of the mucus.

Possible CF treatments for the future

Gene therapy

Current treatments only reduce some of the effects of the disease. Understanding the nature of the gene involved in cystic fibrosis has raised the possibility of a future cure through **gene therapy**. Effective gene therapy would treat the cause rather than the symptoms of the disease, but what is gene therapy?

In gene therapy the genotype and hence the phenotype of target cells (those affected by the disease) is altered. This is achieved as follows:

1   Normal alleles of the gene are inserted into the target cell, either using a genetically modified virus to infect the target cell or using liposomes (spherical phospholipid bilayers).

2   The normal form of the gene is transcribed and translated.

3   A functioning protein is produced in the target cells.

In the case of the CF gene, a functioning CFTR protein is produced and incorporated into the cell membrane, thus restoring the ion channel and avoiding the symptoms of CF.

How genes are inserted using viruses

In the virus, the DNA sequence that allows it to replicate is removed. This is replaced with the normal allele of the desired gene, along with a promoter sequence that initiates transcription and translation of the gene.

When we are infected with some viruses, the viral DNA becomes incorporated into our cells' own DNA. With other viruses the viral DNA remains independent within the nucleus of our cells. In trials with CFTR, the second type of virus is used.

The use of viruses is a potentially efficient form of gene transfer but it has been found to produce an inflammatory response, with patients treated experiencing symptoms such as headache, fatigue, fever and raised heart rate.

How genes are inserted using liposomes

First, a copy of the normal allele is inserted into a loop of DNA (called a **plasmid**). The plasmids are then combined with **liposomes** (spherical phospholipid bilayers). The positively charged head groups of the phospholipids combine with the DNA (a weak acid and so negatively charged) to form a liposome–DNA complex.

The CF patient breathes in an aerosol containing these complexes using a nebuliser. The liposomes fuse with epithelial cell membranes and carry the DNA into the cell (Figure 2.43).

Progress so far

Trials of CF gene therapy started in the UK back in 1993 and the 'normal' CFTR allele has been successfully transferred to the lung epithelial cells of CF patients. The presence of functioning ion channels is indicated by a reduction

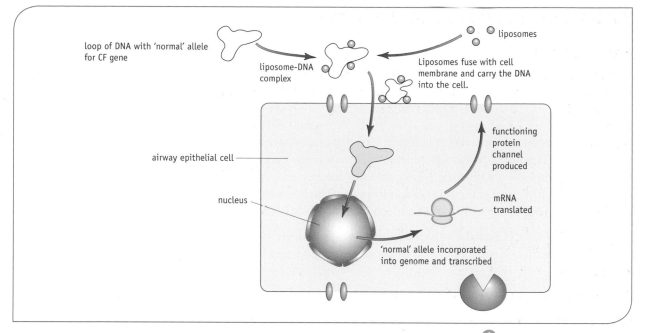

loop of DNA with 'normal' allele for CF gene

liposome-DNA complex

liposomes

Liposomes fuse with cell membrane and carry the DNA into the cell.

functioning protein channel produced

airway epithelial cell

mRNA translated

nucleus

'normal' allele incorporated into genome and transcribed

in the potential difference across the membranes in the nose and lungs, and increased chloride secretions. The results of trials have so far demonstrated a correction of the chloride secretion but not the associated sodium secretion problems.

In trial results published in 1999, chloride transport in the lungs was restored to 25% of normal. However, this type of improvement is temporary. Cells are continuously lost from the epithelium lining the airways so the transfer of the allele to these cells does not offer a permanent solution. The longest the correction has lasted in any trial is about 15 days, so this treatment would have to be repeated throughout the patient's life. Trials continue with the aim of improving the delivery of the gene and thus increasing the effectiveness of the treatment that may one day offer a cure.

The only condition in humans to have been successfully treated using gene therapy by 2004 was a very rare disorder called severe combined immune deficiency (SCID). Patients with this disease cannot make a particular enzyme needed for the immune system to work. In 1990, white blood cells were removed from a four-year-old girl with the disease, Ashanti DeSilva. The alleles for the functioning gene were inserted into her white cells using a virus. The cells were then replaced. Ashanti (now a young woman) has to have regular transfusions with the modified white cells but otherwise is fit and well.

It is hoped that gene therapy will help in the treatment of thalassaemia, haemophilia and sickle cell disease but to date only limited progress has occurred with these diseases. These treatments are all concerned with altering specific **somatic cells** (body cells) and are permitted under UK legislation. The alternative approach of altering the **germ cells** (sperm or eggs) so that every cell in the body contains the new gene is not permitted. There are ethical objections to such **germ line therapy** because of concerns about possible effects in future generations when the new gene is inherited.

Figure 2.43 Gene therapy for cystic fibrosis may one day be successful, but difficulties with the delivery of the gene mean that the trials must continue.

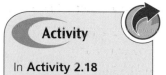

Activity

In **Activity 2.18** complete the exercise on successes and failures in gene therapy. **A2.18S**

2.8 Testing for CF

Even though one person in 25 in the UK is a CF carrier, the first time that most carriers realise that they carry the CF allele is when one of their children is born with the disease.

Once cystic fibrosis is suspected, it is possible to carry out conventional tests to confirm the diagnosis. Ancient European folklore warned that a child who tastes salty when kissed would die young, and this has become the basis of a modern test that measures the level of salt in sweat. The test works because affected people have markedly higher concentrations of salt in their sweat.

People with CF have elevated levels of the protein trypsinogen in their blood. A blood test for this protein is sometimes used in screening newborn babies for the disease.

Early diagnosis of CF allows treatment to begin immediately, which can improve health in later years. Testing of all newborn babies was introduced in Wales in 1997 and in Scotland in January 2003, and there were plans to have all UK babies routinely tested by 2004. At the time of writing (January 2005) about 20% of babies in England were being tested and plans to test all babies were being implemented.

Genetic testing

The gene that codes for the CFTR protein was sequenced in 1989. This led to the possibility of **genetic testing**, identifying the abnormal gene in the DNA of any cells. This in turn has paved the way for **genetic screening** to confirm the results of conventional tests, to identify carriers and also to diagnose CF in an embryo or fetus. (The terms 'genetic testing' and 'genetic screening' are sometimes distinguished but often, as here, used interchangeably.)

How is genetic testing done?

Genetic testing can be performed on any DNA, so it is possible to take samples of cheek cells, white blood cells or cells obtained from a fetus or embryo. The DNA is tested to see whether it contains the known base sequences for the most common mutations that cause cystic fibrosis.

Q2.30 The test is performed only for the most common mutations that cause CF. What does this mean about the reliability of the test? Will there be any false positives or negatives?

Once obtained, the DNA sample is tested by the method illustrated in Figure 2.44 and described below. The essence of this approach applies to many genetic tests.

1 Extracting DNA and cutting it into fragments

DNA extraction removes DNA from the cells. **Restriction enzymes** are used to cut the DNA into fragments.

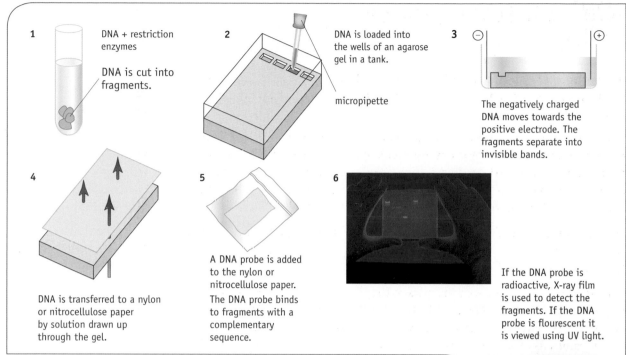

1 DNA + restriction enzymes

DNA is cut into fragments.

2 DNA is loaded into the wells of an agarose gel in a tank.

micropipette

3 The negatively charged DNA moves towards the positive electrode. The fragments separate into invisible bands.

4 DNA is transferred to a nylon or nitrocellulose paper by solution drawn up through the gel.

5 A DNA probe is added to the nylon or nitrocellulose paper. The DNA probe binds to fragments with a complementary sequence.

6 If the DNA probe is radioactive, X-ray film is used to detect the fragments. If the DNA probe is flourescent it is viewed using UV light.

Restriction enzymes, more correctly called restriction endonucleases, are found naturally in bacteria where their function is to cut up invading viral DNA. The value of these restriction enzymes to biologists lies in the fact that they will only cut the DNA at specific base sequences (usually just a few bases long). For example, the enzyme *Eco*RI comes from the bacterium *Escherichia coli* strain RY13 and it splits DNA wherever the sequence of bases is GAATTC. Figure 2.45 shows where it cuts the DNA.

In nature, bacteria are able to protect their own DNA from these enzymes by modifying the bases in the specific sequence targeted by their own restriction enzymes. Restriction enzymes are also used in preparing DNA for gene therapy.

2, 3 Separating the fragments

Gel electrophoresis is used to separate the fragments according to their size. The DNA fragments are placed on a gel made of agarose or polyacrylamide, both of which provide a stable medium through which the DNA molecules can move.

The gel is connected to electrodes that produce an electrical field, and the DNA fragments migrate in the field according to their overall charge and size. Smaller fragments travel faster and therefore further in a given time.

4 Transferring the fragments to paper

Southern blotting is used to transfer the DNA fragments to a nylon or nitrocellulose filter. The filter is placed over the gel and then dry absorbent paper is used to draw the DNA fragments from the electrophoresis gel onto the nylon or nitrocellulose. The paper acts as a wick to draw an alkaline buffer solution through the gel. The DNA strands separate, exposing the base sequences.

▲ **Figure 2.44** The steps in carrying out a genetic test for the CF mutation or many other gene mutations.

▲ **Figure 2.45** *Eco*RI always cuts DNA at this site. Each restriction enzyme cuts at a specific sequence.

5 Adding the gene probe

Gene probes are short sequences of DNA that have a complementary base sequence to the gene being sought. In the case of cystic fibrosis this will be complementary to a known mutant sequence. Traditionally, the probe is made using radioactive phosphorus (^{32}P), making the DNA fragment radioactive. More recent methods use DNA probes that have had fluorescent molecules chemically bound to them. Large quantities of the probe are added to the filter and allowed time to bind with complementary sequences (hybridisation) before any unbound probe is washed away.

6 Determining whether the probe has bound to any DNA fragments

In the case of radioactive probes, the nylon or nitrocellulose filter is dried and placed next to X-ray film. Any radioactive probe present will expose or blacken the film, showing that the DNA sequence being tested for is present. For fluorescent probes, the filter can be viewed under ultraviolet light to reveal the bound DNA.

How can genetic screening be used?

To confirm a diagnosis

Genetic testing can confirm a diagnosis of CF. However, since there are a large number of different mutations of the CFTR gene which cause the disease, a negative result must be treated with caution. It is not currently feasible to test for all of the hundreds of possible mutations that lead to CF.

To identify carriers

Genetic testing can identify carriers. A sample of blood or cells taken from inside the mouth can be used to detect abnormal alleles in people without the disease who are heterozygous. Where there has been a history of cystic fibrosis in a family this can be of value in assessing the probability of having a child with the disease. Counselling is offered before and after testing, and parents can make informed decisions about how to proceed.

For testing embryos

Genetic testing can be used for prenatal testing of an embryo. Currently, there are two techniques used for obtaining a sample of cells from a child before birth. The more common is **amniocentesis**, which involves inserting a needle into the amniotic fluid to collect cells that have fallen off the placenta and fetus (Figure 2.46). This can be carried out at around 15–17 weeks of pregnancy, and involves a risk of causing a miscarriage of between 0.5% and 1%.

Chorionic villus sampling (CVS) is the second technique. Here, a small sample of placental tissue (which includes cells of the fetus) is removed, either through the wall of the abdomen or through the vagina. This can be carried out earlier, between 8 and 12 weeks, since there is no need to wait for amniotic fluid to develop. However, it carries a risk of about 1% to 2% of inducing a miscarriage.

Activities

Use restriction enzymes and try out gel electrophoresis using the simulation in **Activity 2.19**. A2.19S

Or try them out practically in **Activity 2.20**. A2.20S

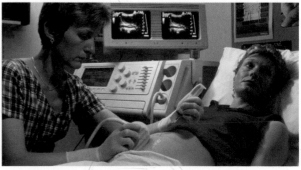

▼ **Figure 2.46** A prenatal test during pregnancy such as amniocentesis (shown here) or CVS involves techniques that carry some risk.

If amniocentesis or CVS results in a positive genetic test for the disease, one possibility is for the woman to have an abortion. Having an abortion is easier for the woman, both physically and emotionally, in the earlier weeks of pregnancy. This can mean that the higher risk of miscarriage associated with CVS is considered worthwhile.

Q2.31   Are there circumstances where people might choose not to test, even if the test was offered to them?

Activity

Activity 2.21 looks at issues involved in using widespread carrier screening. **A2.21S**

Testing before implantation (PIGD)

Genetic testing can enable pre-implantation genetic diagnosis (PIGD). When carrying out *in vitro* fertilisation (IVF), it is possible to test an embryo before it has implanted in the uterus. A cell can be removed from an embryo growing in culture when it has only 8 or 16 cells without harming the embryo (Figure 2.47). The DNA of the cell can be analysed and the results used to decide whether to place the embryo into the womb. IVF, however, is still an expensive and fairly unreliable procedure, and although this avoids the need for a possible abortion, this approach is not used routinely.

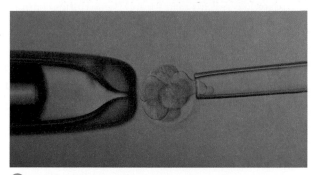

▲ **Figure 2.47** Why can one cell be removed at this stage without harming the developing embryo? We will see the answer to this question in Topic 3. Magnification ×110.

Making ethical decisions – What is right and what is wrong?

How should we decide in life what is right and what is wrong? For example, should we always tell the truth? Can we ever justify turning down a request for help? Should Claire and Nathan have an abortion if they find that their unborn baby has cystic fibrosis?

All of us have **moral** views about these and other matters. For example, you might hold that lying and abortion are always wrong and helping people always right. But in order to maintain that something is ethically acceptable or unacceptable, you must be able to provide a reasonable explanation as to *why* that is the case.

There is no one universally accepted way of deciding whether something is ethically acceptable or not. What there are instead are a number of ethical frameworks, each of which allows you to work out whether a particular action would be right or wrong if you accept the ethical principles on which the framework is based. Usually you get the same answer whichever framework you adopt. But not always! This is why perfectly thoughtful, kind and intelligent people sometimes still disagree completely about whether a particular course of action is justified or not.

Ethical frameworks

We shall examine four widely used ethical frameworks. You should find these of value when considering various issues, such as genetic screening and abortion, raised in this topic. We will also refer to these frameworks in other topics in this course.

1 Rights and duties

Most of us tend to feel that there are certain human **rights** that should always be permitted. For example, we talk about the right to life, the right to a fair trial and the right to freedom of speech. Certain countries, for example the USA, have some of these rights enshrined in their constitutions.

If you have a right to something, then I may have particular **duties** towards you. For example, suppose that you are a 6-month-old baby with a right to life and I am your parent. I have a duty to feed you, wash you, keep you warm and so on. If I don't fulfill these duties, I am failing to carry out my responsibilities and the police or social services may intervene.

But where do rights come from? Some people with a religious faith find them in the teachings of their religion. For example, the ten commandments in the Jewish scriptures talk about not stealing, not murdering, telling the truth and so on.

But nowadays, of course, many people, indeed in the UK most people, have little or no religious faith. So where can they – perhaps you – find rights? The simplest answer is that rights are social conventions built up over thousands of years. If you want to live in a society you have, more or less, to abide by its conventions.

2 Maximising the amount of good in the world

Perhaps the simplest ethical framework says that each of us should do whatever *maximises the amount of good in the world*. For example, should I tell the truth? Usually yes, as telling lies often ends up making people unhappy and unhappiness is not good. But sometimes telling the truth can lead to more unhappiness. If your friend asks you if you like the present they have just given you and you don't, would you tell the truth? Most of us would tell a 'white lie', not wanting to hurt their feelings.

This ethical approach is known as **utilitarianism**. Notice that utilitarians have no moral absolutes beyond maximising the amount of good in the world. A utilitarian would hesitate to state that anything is always right or always wrong. There might be circumstances in which something normally right (e.g. keeping a promise) would be wrong and there might be circumstances when something normally wrong (e.g. killing someone) would be right.

3 Making decisions for yourself

One of the key things about being a human is that we can make our own decisions. There was, for example, a time when doctors simply told their patients what was best for them. Now, though, there has been a strong move towards enabling patients to act autonomously. People act autonomously when they make up their own mind about something. If you have ever had an operation, you will probably have signed a consent form. The thinking behind this is that it isn't good for a surgeon to be allowed to operate on you unless you have given **informed consent**.

> ### Checkpoint ✓
>
> **2.7** Use each of these four ethical frameworks to consider whether or not it is acceptable to abort a fetus found by amniocentesis to have CF.

Of course, it is perfectly possible autonomously to decide to be absolutely selfish! A utilitarian would say we need to weigh the benefits of someone acting autonomously with any costs of them doing so. Only if the overall benefits are greater than the overall costs is **autonomy** desirable. Someone who believes in rights and duties might say that each of us has a right to act autonomously but also has a duty to take account of the effects of our actions on others.

4 Leading a virtuous life

A final approach is one of the oldest. This holds that the good life (in every sense of the term) consists of acting virtuously. This may sound rather old fashioned but consider the **virtues** that you might wish a good teacher/ lecturer to have. She or he might be understanding, be able to get you to learn what you want or need to learn, and believe in treating students fairly.

Traditionally the seven virtues were said to be **justice**, prudence (i.e. wisdom), temperance (i.e. acting in moderation), fortitude (i.e. courage), faith, hope and charity. Precisely what leading a virtuous life means can vary and isn't always straightforward. Think about the virtues you might like to see in a parent, a doctor and a girl-/boyfriend. What would be the virtuous course of action for Claire and Nathan?

Genetic counselling

If a couple like Nathan and Claire are at risk of having a child with a genetic disorder, a genetic counsellor can provide advice. Such a person will help the couple understand how the disease is inherited and the chance that any child they conceive will have the disease. A genetic counsellor will explain the tests available and the possible courses of action, depending on the outcome of the tests. The counselling should help the couple decide such things as whether to be tested and whether to have children, use *in vitro* fertilisation with screening or use prenatal screening.

Q2.32 If both parents are found to be carriers, what are the options open to them?

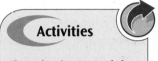

Activities

The role play in **Activity 2.22** lets you think about some of the issues covered in this topic. **A2.22S**

Use **Activity 2.23** to check your notes using the topic summary provided. **A2.23S**

Topic test

Now that you have finished Topic 2, complete the end-of-topic test before starting Topic 3.

Topic ③ Voice of the genome

Why a topic called Voice of the genome?

Most organisms, from apple trees to zebras, including llamas, legumes, mice, monkeys, mushrooms, mayflies and us, *Homo sapiens*, start out in much the same way – as a very simple undifferentiated cell. But from that cell come scores of different cells, specialised for a huge variety of jobs and all produced in exactly the correct place. How is the amazing change from a single cell into a complex moving and communicating body achieved (Figure 3.1)? In this topic we follow the fate of the gametes produced by parents as these gametes start on the greatest journey of all – from separate cells to a full-grown adult. We discover how this development process is controlled and what can go wrong when the control stops working properly.

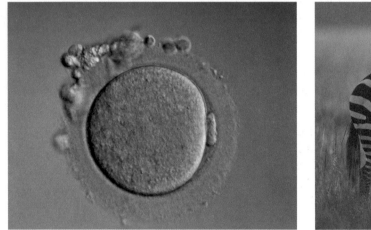

▲ **Figure 3.1** From a single egg to a full-grown adult.

William Harvey (1578–1657), one of the greatest anatomists who ever lived, said 'Everything comes from eggs'. But people used to think that the whole organism was wrapped up in miniature inside the sperm, ready to grow and be unfolded during development (Figure 3.2). Of course this isn't the case, but the truth is not so different. Inside every cell is a complete genome – DNA containing a full set of genes that control the growth and development of the whole organism.

We share 98.5% of our DNA sequence with chimpanzees, making them our closest relatives in the animal kingdom. However, we also share 50% of our genes with bananas, which is food for thought!

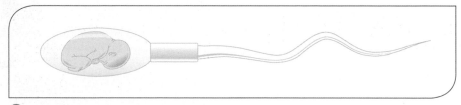

🔺 **Figure 3.2** It was once thought that a miniature baby was to be found inside the sperm cell.

In humans there are known to be approximately 35 000 genes. How do these genes talk to the cell? How do they provide the instruction manual to build and control not only a single cell but the great range of specialised cells that make up an adult organism? How do cells provide instructions like 'Grow a leg here and do it now'?

In some organisms, features you might think are controlled exclusively by genes, such as fur colour, are not quite so straightforward, with genes and the environment interacting. How do genes and the environment interact? Is cancer inherited, caused by our environment, or a combination of both?

The year 2000 saw the publication of the draft DNA sequence for the human genome. The rapidly advancing knowledge about our genes and how they work may be both a blessing and a curse. Here we look at how the Human Genome Project has opened the door to both great advances and ethical dilemmas.

Overview of the biological principles covered in this topic

Gamete structure and function provide the starting point for this topic. You will look at their role in fertilisation to form the zygote. To understand how a single cell divides and grows into a whole organism, you need a good grasp of cell structure and a detailed knowledge of the cell cycle. At GCSE the basics of cell division were covered; here you will gain a more detailed picture of this process, including the role of both mitosis and cytoplasmic division.

You will discover how the first cells produced in the early human embryo have the potential to develop into any part of the body, and see how these stem cells are becoming increasingly important to medical research. They offer a huge potential for the creation of new treatments and therapies. But these new developments are not welcomed by all, and you will have to consider the arguments for and against the use of these stem cells in medicine to clarify your own views on the issue.

Building on the basic idea of cells becoming specialised and working together as tissues, organs and organ systems, you will see how this development is made possible at the level of the individual cell. You will discover how gene expression is regulated and how this contributes to controlling development.

Review

Are you ready to tackle Topic 3 *Voice of the genome?*

Complete the GCSE review and GCSE review test before you start.

3.1 In the beginning

Have you ever tried incubating an egg bought from the supermarket (Figure 3.3)? If so it is unlikely that the egg hatched because these eggs are unfertilised. The hens they come from are kept in female-only flocks; no cockerels are present so no sperm have combined with these eggs. To get a chick you need a fertilised egg.

The life cycle of most organisms starts in this way. Gametes combine to form a single fertilised ovum that divides and differentiates into a complex body. You will remember from GCSE that the **sperm** and **ovum** (egg) cells are the **gametes** or sex cells. They are shown on page 102, in Figure 3.11.

▲ **Figure 3.3** Which came first?

Key biological principle: Are all cells basically the same?

Cells do have many common features, but it is possible to distinguish two basic types: **prokaryotic** and **eukaryotic** cells.

Prokaryotic cells

Bacteria and cyanobacteria (photosynthetic bacteria) together make up the Prokaryotae Kingdom. Their cells do not have nuclei or other membrane-bound cell organelles (Figures 3.4 and 3.5). This type of cell is called a 'prokaryotic cell', meaning 'before the nucleus'. Most prokaryotes are extremely small with diameters between 0.5 and 5 μm. Their DNA is not associated with any proteins and lies free in the cytoplasm. A cell wall is always present in prokaryotic cells.

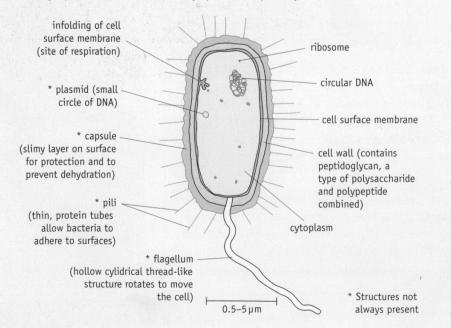

infolding of cell surface membrane (site of respiration)

ribosome

circular DNA

* plasmid (small circle of DNA)

cell surface membrane

* capsule (slimy layer on surface for protection and to prevent dehydration)

cell wall (contains peptidoglycan, a type of polysaccharide and polypeptide combined)

* pili (thin, protein tubes allow bacteria to adhere to surfaces)

cytoplasm

* flagellum (hollow cylidrical thread-like structure rotates to move the cell)

0.5–5 μm

* Structures not always present

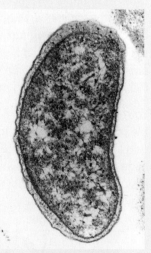

▲ **Figure 3.5** Electron micrograph of the bacterium *Vibrio cholerae*. Magnification ×36 000.

◄ **Figure 3.4** The structure of a generalised bacterial cell.

Eukaryotic cells

All other living organisms have cells that contain discrete membrane-bound organelles such as nuclei, mitochondria and chloroplasts (plants only). These are eukaryotic cells, meaning 'true nucleus'. Eukaryotic cells are larger than prokaryotic cells with diameters of 20 μm or more. Unlike the prokaryotes, not all eukaryotes have a cell wall. Organisms with eukaryotic cells are eukaryotes and are grouped into four other kingdoms – the animal, plant, fungus, and protoctist kingdoms. The protoctist kingdom contains the various unicellular eukaryotic organisms, such as *Amoeba*, as well as the seaweeds.

Inside the eukaryotic animal cell

The classical children's drawing of a cell resembles a fried egg: one circle with a large black blob in the middle. Using a simple light microscope this is almost as much detail as it is possible to see. However, the use of electron microscopy reveals a wealth of fine structures within the cell, often described as the cell ultrastructure (Figure 3.6). Figure 3.7 shows a two-dimensional representation of the cross-section of a generalised animal cell. You can see there are many structures within the cell cytoplasm. These are shown in three dimensions in Figure 3.8. Use these diagrams to answer the questions below.

Q3.1 Follow a straight line across the cell in Figure 3.7 from A to B and identify each of the structures that the line crosses using Figure 3.8 on page 100.

Q3.2 Identify two structures within the cell which are not crossed by this line.

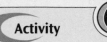

Activity

In **Activity 3.1** you look in more detail at the relationship between the three-dimensional structure and function of organelles in the cell using the interactive cell. **A3.01S**

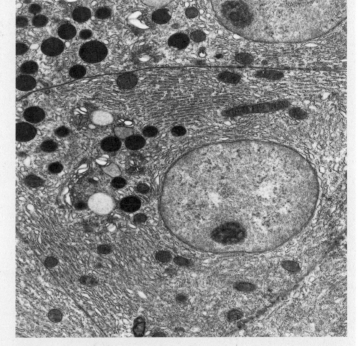

Figure 3.6 Electron micrograph of a pancreas cell shows the cell's ultrastructure. It contains many mature secretory vesicles (dark round structures) and immature secretory vesicles (paler round structures).

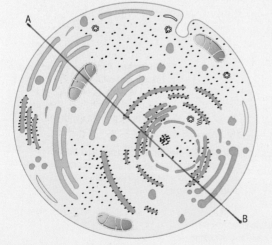

Figure 3.7 A generalised animal cell.

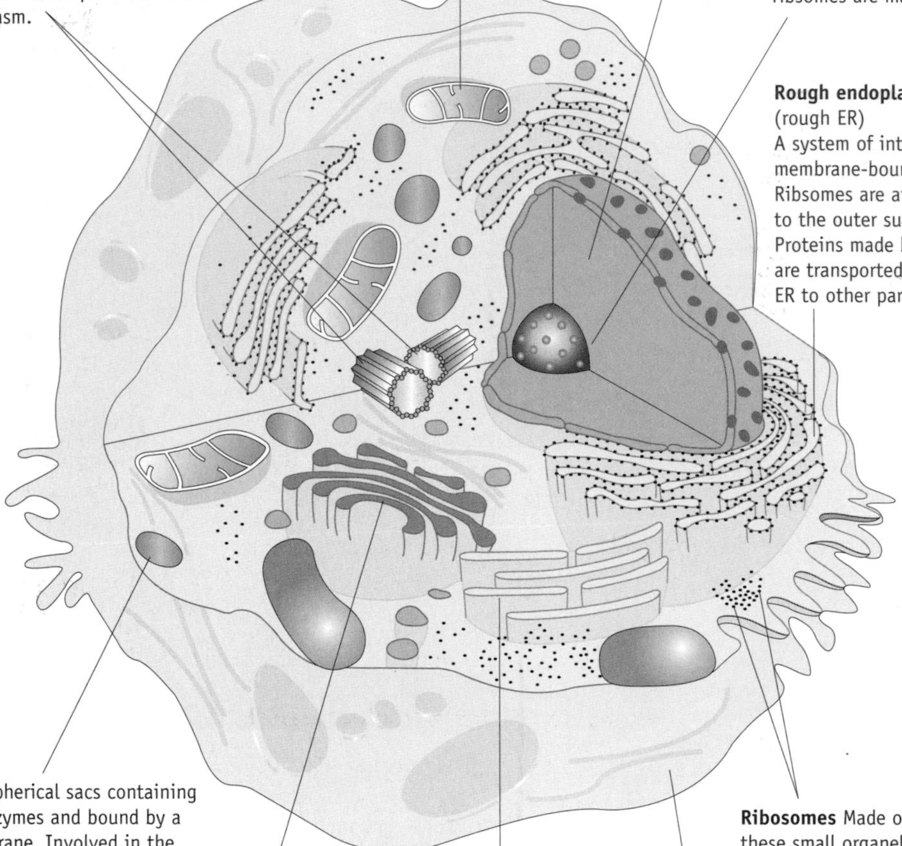

Mitochondrion (plural mitochondria) The inner of its two membranes is folded to form finger-like projections called cristae. The mitochondria are the site of the later stages of aerobic respiration.

nucleus Enclosed by an envelope composed of two membranes perforated by pores. Contains chromosomes and a nucleolus. The DNA in chromosomes contains genes that control the synthesis of proteins.

Centrioles Every animal cell has one pair of centrioles, which are hollow cylinders made up of a ring of nine protein microtubules (polymers of globular proteins arranged in a helix to form a hollow tube). They are involved in the formation of the spindle during nuclear division and in transport within the cell cytoplasm.

The nucleolus is a dense body within the nucleus and is where ribosomes are made.

Rough endoplasmic reticulum (rough ER) A system of interconnected membrane-bound, flattened sacs. Ribosomes are attached to the outer surface. Proteins made by these ribosomes are transported through the ER to other parts of the cell.

Lysosome Spherical sacs containing digestive enzymes and bound by a single membrane. Involved in the breakdown of unwanted structures within the cell, and in destruction of whole cells when old cells are to be replaced or during development. The acrsome is a specialised lysosome.

Ribosomes Made of RNA and protein, these small organelles are found free in the cytoplasm or attached to endoplasmic reticulum. They are the site of protein synthesis.

Cell surface membrane (plasma membrane) Phospholipid bilayer containing proteins and other molecules forming a partially permeable barrier.

Golgi apparatus Stacks of flattened, membrane-bound sacs formed by fusion of vesicles from the ER (Figure 3.9). Modifies proteins and packages them in vesicles

Smooth endoplasmic reticulum (smooth ER) Like rough ER, but does not have any attached ribosomes. Smooth ER makes lipids and steroids

▲ **Figure 3.8** A three-dimensional representation of a generalised animal cell.

The structure and function of plant cells will be studied in Topic 4.

Cells are dynamic

The cell is not static; there is continual movement of molecules within it. In Topic 2 you saw the movement of RNA from the nucleus to the ribosomes for protein synthesis. Once the proteins have been synthesised they are processed and move through the cell to where they are needed. In addition, many proteins such as enzymes, hormones and signal proteins are released from cells. This trafficking of proteins through the cell involves the endoplasmic reticulum, Golgi apparatus and vesicles. These membrane structures are continually created and lost within the cell, as outlined in Figure 3.9.

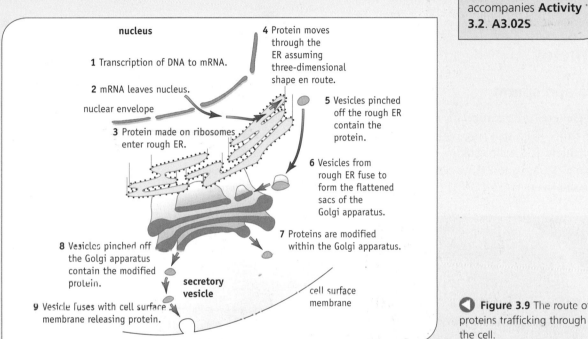

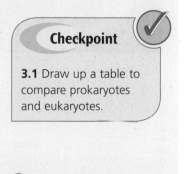

Activity

You can see the protein trafficking within the cell in the animation that accompanies **Activity 3.2**. **A3.02S**

Figure 3.9 The route of proteins trafficking through the cell.

Q3.3 Look at the electron micrograph in Figure 3.10 below and identify the structures labelled **A** to **E**.

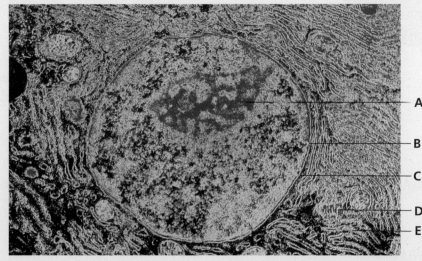

Checkpoint

3.1 Draw up a table to compare prokaryotes and eukaryotes.

Figure 3.10 Electron micrograph of part of a cell from the pancreas of the bat *Myotis lucifugus*. The cell produces pancreatic juice and enzymes. Magnification ×1300.

Did you know? The origins of chloroplasts and mitochondria

Chloroplasts and mitochondria are thought to have originally been prokaryotic cells. Very early in the evolution of multicellular organisms, they developed a mutualistic relationship (of benefit to both partners) with eukaryotic cells. Perhaps these ancestors of chloroplasts and mitochondria gained protection and a more constant environment by being inside eukaryotes. The eukaryotes probably benefited from the products of the metabolism of the prokaryotes:

sugar and oxygen from the photosynthetic prokaryotes that became chloroplasts, and ATP from the non-photosynthetic prokaryotes that became mitochondria. To this day, both chloroplasts and mitochondria contain their own DNA separate from the DNA of the cells they inhabit. However, they can no longer replicate themselves independently but need to rely partly on enzymes made by their hosts.

Gametes

Ova and sperm

The gametes are adapted for their roles in sexual reproduction. In humans, as in other mammals, the ovum is a large cell incapable of independent movement. It is wafted along one of the oviducts from the ovary to the uterus (Figure 2.26A, page 73) by ciliated cells lining the tubes and by muscular contractions of the tubes. The cytoplasm of the ovum contains protein and

Checkpoint

3.2 Look at Figure 3.11. Make a list of the differences between the gamete cells, and explain the value of any feature.

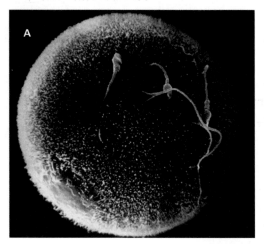

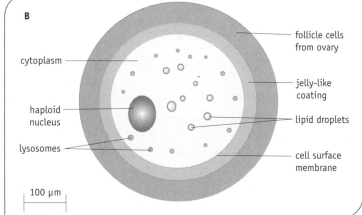

cytoplasm

haploid nucleus

lysosomes

100 µm

follicle cells from ovary

jelly-like coating

lipid droplets

cell surface membrane

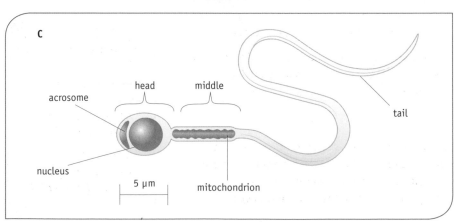

acrosome

head   middle

nucleus

5 µm   mitochondrion

tail

▲ **Figure 3.11 A** Scanning electron micrograph of sperm on an ovum during fertilisation. Magnification ×1100. Diagrams of **B** a human ovum and **C** a human sperm.

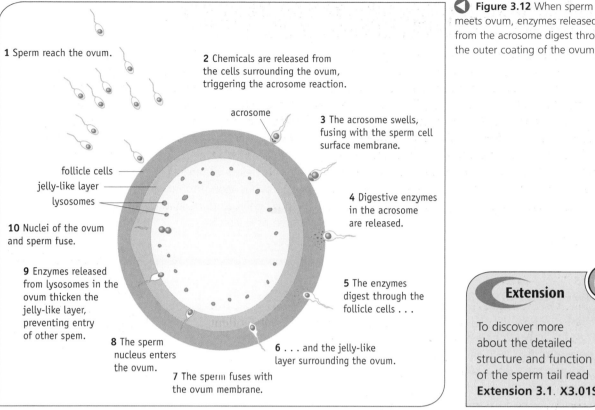

Figure 3.12 When sperm meets ovum, enzymes released from the acrosome digest through the outer coating of the ovum.

1 Sperm reach the ovum.

2 Chemicals are released from the cells surrounding the ovum, triggering the acrosome reaction.

acrosome

3 The acrosome swells, fusing with the sperm cell surface membrane.

follicle cells

jelly-like layer

lysosomes

4 Digestive enzymes in the acrosome are released.

10 Nuclei of the ovum and sperm fuse.

9 Enzymes released from lysosomes in the ovum thicken the jelly-like layer, preventing entry of other spem.

5 The enzymes digest through the follicle cells . . .

8 The sperm nucleus enters the ovum.

6 . . . and the jelly-like layer surrounding the ovum.

7 The sperm fuses with the ovum membrane.

lipid food reserves for the early development of the embryo. Surrounding the cell is a jelly-like coating called the zona pellucida.

The sperm cell is much smaller than the ovum and is motile (can move). To enable it to swim, the sperm cell has a long tail powered by energy released by mitochondria. Men continually produce large numbers of sperm once they have reached puberty. If sperm enter the vagina they swim through the uterus, their passage assisted by muscular contractions of the uterus walls. Once in the oviducts they may meet the ovum if intercourse takes place at about the time of ovulation. The sperm are attracted to the ovum by chemicals released from the egg. To penetrate the ovum, the **acrosome** in the head of the sperm releases digestive enzymes, which break down the jelly-like coating of the ovum (Figure 3.12). The acrosome is a type of lysosome. Lysosomes are enzyme-filled sacs found in the cytoplasm of many cells. They are involved in the breakdown of unwanted structures within the cell.

Once a sperm fuses with and penetrates the membrane surrounding the egg, chemicals released by the ovum cause the jelly-like layer to thicken, preventing any further sperm entering the egg.

Getting together

To produce a new individual, the nuclei from the gametes have to combine in the process of **fertilisation**. In fertilisation the nucleus from one sperm enters the ovum, and the genetic material of the ovum and sperm fuse, forming a fertilised ovum called a **zygote**. This cell now contains genetic material from both parents.

Extension

To discover more about the detailed structure and function of the sperm tail read **Extension 3.1**. **X3.01S**

Activities

In **Activity 3.3** you can relate the structure of gametes to their functions. **A3.03S**

You can watch fertilisation in marine worms in **Activity 3.4**.

For most organisms the diploid cell formed at fertilisation is just the starting point on the road to creating a more complex multicellular structure. This single cell will divide and give rise to numerous more specialised cells, creating the huge variety of structures within a body.

Did you know? One cause of male infertility

For the human zygote to develop, the gamete nuclei have to fuse and a chemical from the sperm cytoplasm is required to activate the fertilised cell. This chemical is a protein called oscillin. It causes calcium ions to move in and out of stores in the cytoplasm of the ovum. These oscillations of calcium ion concentration trigger the zygote to begin developing into an embryo. Oscillin is concentrated in the first part of the sperm to attach to the ovum, and enters before the male nucleus so as to activate the ovum. It is thought that low levels of oscillin in sperm may be linked to male infertility and this is a current area of research.

Gamete cells are unusual

Although the number of chromosomes varies from species to species, all 'normal' individuals within a species will have the same number of chromosomes in each cell. Human cells contain 46 chromosomes made up of 22 homologous pairs and one pair of sex chromosomes; fruit fly cells contain eight chromosomes made up of three homologous pairs and one pair of sex chromosomes; and so on.

Extension

To find out about sperm competition read **Extension 3.2. X3.02S**

The fundamental difference between gametes and other cells is the number of chromosomes they contain. In Topic 2 you saw that gametes have half the number of chromosomes found in normal cells – one chromosome from each homologous pair – but have you considered why this is? Think about what happens at fertilisation. If the sperm and ovum cells both had the full chromosome number, such as 46 in humans, then the zygote would have 92 chromosomes. When this individual reproduced the zygote would contain 184 chromosomes. With each generation the number of chromosomes would double, which obviously wouldn't work. This is why the gametes contain half the full number of chromosomes, that is 23 in humans, made up of one of each homologous pair and one sex chromosome. When the gametes fuse, the full number of 46 is restored and each chromosome can pair up again, as you can see in Figure 3.13.

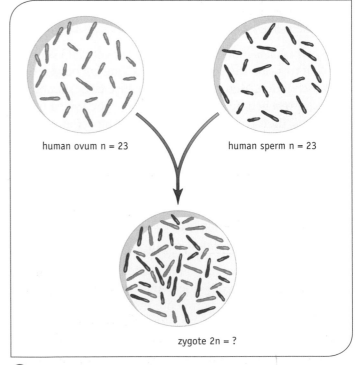

human ovum n = 23   human sperm n = 23

zygote 2n = ?

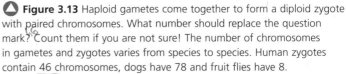

🔺 **Figure 3.13** Haploid gametes come together to form a diploid zygote with paired chromosomes. What number should replace the question mark? Count them if you are not sure! The number of chromosomes in gametes and zygotes varies from species to species. Human zygotes contain 46 chromosomes, dogs have 78 and fruit flies have 8.

How do gametes form?

There are two different types of cell division in living organisms. One produces new body cells as an organism grows and develops. This retains the full number of chromosomes, called the **diploid** number (2*n*), 46 in humans, and is called **mitosis**. The other type of cell division produces the gametes. They have only half the number of chromosomes, called the **haploid** number (*n*), 23 in humans. This reductive form of cell division occurs in the ovaries and testes of animals, and the ovaries, anthers or other gamete-producing structures in plants. This type of cell division is called **meiosis** and is shown in Figure 3.14.

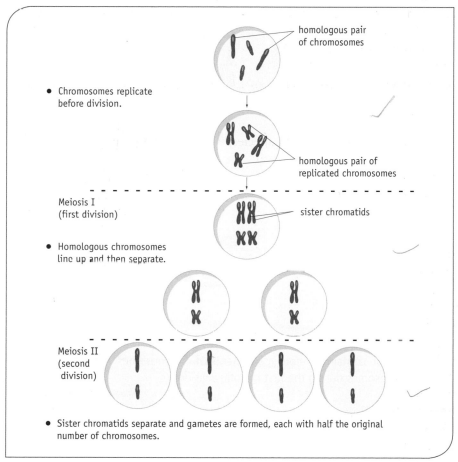

- Chromosomes replicate before division.

homologous pair of chromosomes

homologous pair of replicated chromosomes

Meiosis I (first division)

sister chromatids

- Homologous chromosomes line up and then separate.

Meiosis II (second division)

- Sister chromatids separate and gametes are formed, each with half the original number of chromosomes.

Figure 3.14 Gamete production by meiosis.

Meiosis has two important roles in biology. First it results in haploid cells, which are necessary to maintain the diploid number after fertilisation. Secondly, it helps create genetic variation among offspring.

How does meiosis result in genetic variation?

During the process of meiosis, only one chromosome from each homologous pair ends up in each gamete. This process is random; either chromosome from each pair could be in any gamete. The way that this happens is shown in Figure 3.15. If you consider an organism with six chromosomes, that is three homologous pairs **AA**, **BB** and **CC**, it could form eight combinations in its gametes.

AA   BB   CC
AB   BC
AC

This way of sharing out chromosomes, called **random assortment**, produces genetically variable gametes. When these join with another set at fertilisation, this pretty well guarantees that individuals produced from sexual reproduction are genetically different from each other.

Activity

In **Activity 3.5** you can model chromosome assortment. **A3.05S**

The arrangement of each chromosome pair during the first division in meiosis is completely random. In a cell with three chromosome pairs both the arrangements shown here are possible, for example.

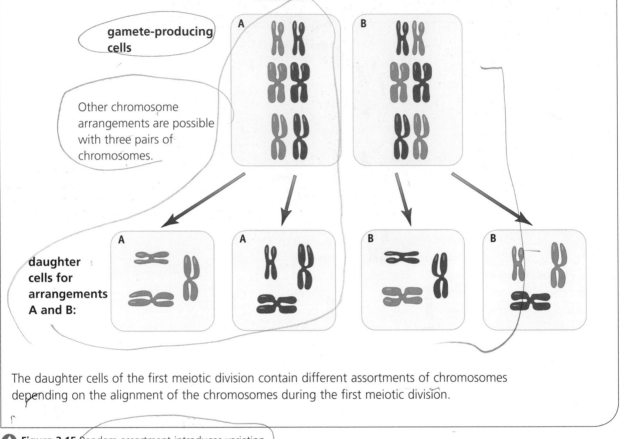

gamete-producing cells

Other chromosome arrangements are possible with three pairs of chromosomes.

daughter cells for arrangements A and B:

The daughter cells of the first meiotic division contain different assortments of chromosomes depending on the alignment of the chromosomes during the first meiotic division.

△ **Figure 3.15** Random assortment introduces variation.

Human cells contain 23 pairs of chromosomes. This random assortment of the chromosome pairs as they line up during meiosis I is a source of genetic variation. The number of possible combinations of the chromosomes is so large that it is unlikely that two siblings will ever have the same genetic make-up unless they are identical twins.

Q3.4 How many possible combinations of maternal and paternal chromosomes could be found in the gametes of organisms with $2n = 8$, and organisms with $2n = 10$?

3.2 From one to many: the cell cycle

3.2 From one to many: the cell cycle

Figure 3.16 How many cells are there in an elephant? Humans have approximately ten thousand thousand million, 10^{13}, cells.

Elephants (Figure 3.16) have much bigger bodies than mice. However, elephant cells are little or no larger than the corresponding mouse cells – there are just far more of them. Elephants and mice both start as a single fertilised cell, the zygote, with the potential to divide and grow into the complete body of a new individual. To make the vast number of cells required to build individuals, new cell contents must be synthesised and then one cell must divide into two. This well-organised pattern of events is called the **cell cycle**.

The cell cycle

The cell cycle can be divided into two distinct parts: interphase and division (Figure 3.17).

Preparation for division: interphase

Interphase is a time of intense and organised activity as the cell synthesises new cell components such as organelles and membranes, and new DNA. The formation of new cellular proteins occurs throughout interphase, whereas DNA synthesis occurs during the **S** (for synthesis of DNA) phase in the middle of interphase. The S phase separates the first gap or **G1** phase from the second gap or **G2** phase, as shown in Figure 3.17.

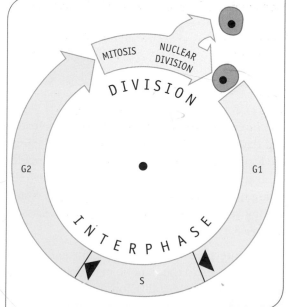

Figure 3.17 The stages in the cell cycle.

The length of interphase differs depending on the role of the cells. In the developing human embryo, there is no interphase for the first few divisions – the zygote already contains the materials needed to form the first 16 or so cells. In these first few divisions, the embryo divides without growing in size, producing smaller cells with each cell cycle. This makes the embryonic cell cycle much faster than those of other body cells.

The S and G2 phases of most cells remain relatively constant in duration. The length of the G1 phase is more variable; some cells can take weeks, months or even years to complete this phase. For example, liver cells may divide only once every one or two years due to an extended G1 phase. Some cells such as nerve and muscle cells never divide again, remaining permanently in a non-dividing state.

Q3.5 Suppose the cell cycle shown in Figure 3.17 lasts approximately 24 hours. Calculate the approximate length of time that the G1, S, G2 and division phases will last.

Look at Figure 3.18. What do you notice about the nucleus during interphase? Can you see individual chromosomes? What does this suggest to you about what may be going on inside the cell?

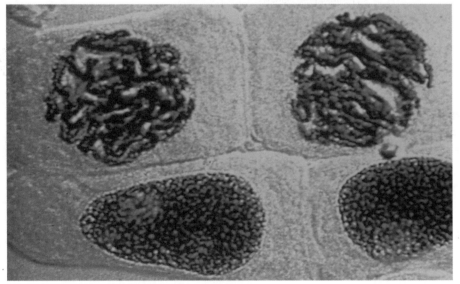

◀ **Figure 3.18** The lower two cells are typical of interphase. Magnification ×6700.

The interphase nucleus is a fairly uniform, featureless structure with one or two darker-staining regions called nucleoli (singular nucleolus). Ribosomes are formed in the nucleoli. Ribosomes are made of protein and rRNA (ribosomal RNA) and appear dark when viewed with an electron microscope, giving nucleoli their dark appearance. The rest of the nucleus contains the chromosomes, which are made up of DNA associated with protein (Figure 3.19).

During interphase, the individual chromosomes are unravelled (Figure 3.19) to allow access to the genetic material so new proteins can be synthesised. In preparation for cell division the cell synthesises additional cytoplasmic proteins and organelles. The cell must also produce copies of DNA for the two

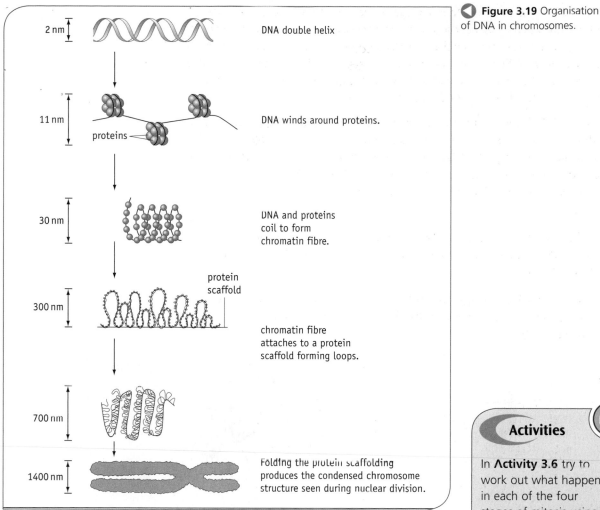

Figure 3.19 Organisation of DNA in chromosomes.

Within the figure:

- 2 nm — DNA double helix
- 11 nm — DNA winds around proteins. (proteins)
- 30 nm — DNA and proteins coil to form chromatin fibre.
- 300 nm — protein scaffold / chromatin fibre attaches to a protein scaffold forming loops.
- 700 nm
- 1400 nm — Folding the protein scaffolding produces the condensed chromosome structure seen during nuclear division.

new cells. It is vital that this DNA is identical in both structure and quantity to the DNA in the original cell. This is achieved by DNA replication, as described in Topic 2, page 81.

Q3.6 Which organelles would you expect to be most active in a cell that is producing large quantities of new proteins?

Cell division

By the end of interphase, the cell contains enough cytoplasm, organelles and DNA to form two new cells. The next step is to share out both the DNA and the contents of the cytoplasm so that each new cell can function independently. The DNA is separated in **nuclear division** (mitosis). **Cytoplasmic division** follows this.

Cell division is a continuous process that moves without pausing, from a single cell with double the usual amount of contents to two new cells. However, it is possible to describe four stages during mitosis by the behaviour of the chromosomes and other structures within the cell. Figure 3.20 shows what is happening in each of these stages, known as **prophase**, **metaphase**, **anaphase** and **telophase**.

Activities

In **Activity 3.6** try to work out what happens in each of the four stages of mitosis using the mitosis flick book. **A3.06S**

Compare this with the cell cycle/mitosis animation in **Activity 3.7**. **A3.07S**

Activities

View the different stages of mitosis in root tip squashes and determine the length of each stage in **Activity 3.8**. **A3.08S**

You can also count cells at different stages using the web tutorial in **Activity 3.9**. **A3.09S**

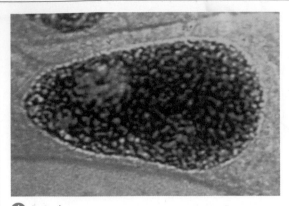

△ Interphase.

During **interphase**, new cell organelles are synthesised and DNA replication occurs. By the end of interphase, the cell contains enough cell contents to produce two new cells.

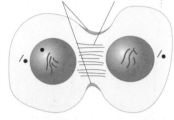

protein filaments involved in cytoplasmic division   microtubules involved in cytoplasmic division

After nuclear division, the final reorganisation into two new cells occurs. This is called **cytoplasmic division**. In animal cells, the cell surface membrane constricts around the centre of the cell. This narrows until the cell is cleaved into two new cells. A ring of protein filaments bound to the inside surface of the cell surface membrane is thought to contract in this process. It has been proposed that the same proteins, actin and myosin, that are responsible for muscle contraction may be the proteins responsible for cell cleavage. Plant cells synthesise a new cell plate between the two new cells instead of showing this cell cleavage (Figure 3.21).

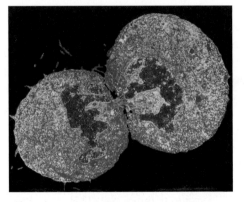

△ Cytoplasmic division.

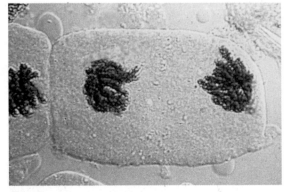

△ 4 Telophase.

Nuclear envelope reforms.

Chromosomes unravel.

This last stage of mitotic division is called **telophase**. This is effectively the reverse of prophase. The chromosomes unravel and the nuclear envelope reforms, so that the two sets of genetic information become enclosed in separate nuclei.

△ **Figure 3.20** Follow the sequence of events during cell division, clockwise in the direction of the arrows. The diagrams show a generalised animal cell containing only four chromosomes.

lecsapódik

During **prophase**, the chromosomes condense, becoming shorter and thicker, with each chromosome visible as two strands called **chromatids**. Apart from the occasional mutation the two strands are identical copies of one another, produced by replication. They are effectively two chromosomes joined at one region called the **centromere** (Figure 3.22).

Chromosomes condense.

3D

Spindle forms.

Nuclear envelope breaks down forming vesicles in the cytoplasm.

Q3.7 Why do the chromosomes condense?

▲ **1** Prophase.

In the next stage of mitosis, **anaphase**, the centromeres split. The spindle fibres shorten, pulling the two halves of each centromere in opposite directions. One chromatid of each chromosome is pulled to each of the poles. Anaphase ends when the separated chromatids reach the poles and the spindle breaks down.

During prophase, microtubules from the cytoplasm form a three-dimensional structure called the **spindle**. The **centrioles** move around the nuclear envelope and position themselves at opposite sides of the cell. These form the poles of the spindle, and are involved in the organisation of the spindle fibres. The spindle fibres form between the poles. The widest part of the spindle is called the equator.

Q3.8 Why does the nuclear envelope disintegrate?

The breakdown of the nuclear envelope signals the end of prophase and the start of **metaphase**. The chromosomes' centromeres attach to spindle fibres at the equator. When this has been completed the cell has reached the end of metaphase.

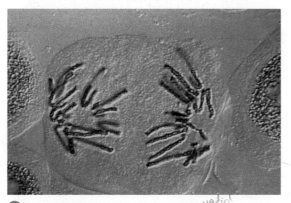

▲ **3** Late anaphase.

chromatid

pole

Spindle fibres shorten, pulling chromatids towards the poles.

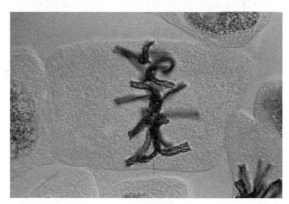

▲ **2** Metaphase.

Chromosomes move to equator.

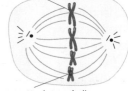

Centromeres attach to spindles.

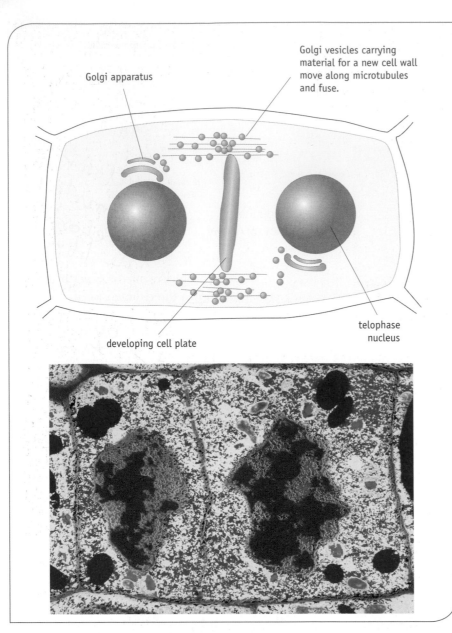

Golgi apparatus

Golgi vesicles carrying material for a new cell wall move along microtubules and fuse.

developing cell plate

telophase nucleus

◀ **Figure 3.21** A plant cell undergoing cytoplasmic division. Photo magnification ×4000.

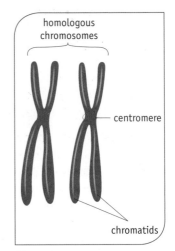

homologous chromosomes

centromere

chromatids

▲ **Figure 3.22** As the chromosomes condense during prophase, each chromosome becomes visible as two strands joined at one point called the centromere. The centromere may be in the middle of the strands or towards one end. Each strand is called a chromatid. The two chromatids are often called sister chromatids.

Q3.9 Look at Figure 3.23 and compare the numbers of cells in telophase with the number in anaphase. What does this suggest to you about the relative lengths of time that these two phases take?

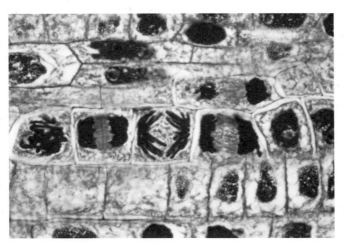

▶ **Figure 3.23** Cells undergoing mitosis. Magnification ×1450.

Checkpoint

3.3 a Draw up a sequence of annotated cells which clearly show what happens during each stage of mitosis.
b Highlight the key features of mitosis that ensure identical daughter cells are produced.

Key biological principle: Why is mitosis so important?

Mitosis ensures genetic consistency, with daughter cells genetically identical to each other and to the parent cell, each containing exactly the same number and type of chromosomes as their parent cell. This is achieved by:

- DNA replication prior to nuclear division
- the arrangement of the chromosomes on the spindle and their separation to the poles.

This genetic stability is important in growth and repair, and also in asexual reproduction.

Growth and repair

Mitosis occurs in the growth of any organism as it develops from a single cell into a multicellular organism. It ensures that a multicellular organism has genetic stability; all the cells in the body have the same genetic information. Some organisms can regenerate lost or damaged parts of their body, and starfish can even regrow a completely new body from a fragment (Figure 3.24A). Mitosis also allows old and damaged cells to be replaced with identical new copies.

Asexual reproduction

Many organisms can reproduce without producing gametes. They grow copies of themselves by mitosis, producing offspring that are genetically identical to each other and to their parent. This form of reproduction is asexual, meaning 'without sex'. Asexual reproduction occurs when bacteria undergo binary fission; the bacterial cell grows and then divides into two new cells. *Hydra* reproduce asexually by budding (Figure 3.24B). Asexual reproduction is common in plants where it is sometimes known as vegetative reproduction. The growth of new plants from tubers and bulbs are familiar examples.

It is worth noting that there are some organisms, such as mosses and liverworts, that reproduce asexually at one stage in their life cycle.

Q3.10 For each pair of statements, decide which one refers to asexual and which to sexual reproduction:

a one parent needed / two parents needed

b gametes / no gametes

c fertilisation / no fertilisation

d offspring show genetic variation / offspring genetically identical

e allows for rapid increase in population numbers / allows only slow increase in numbers

f unfavourable changes in the environment kill whole population / changes in the environment only kill poorly adapted individuals

g effective for rapid colonisation of new areas / enables adaptation to changes in the environment.

Extension

To find out more about asexual reproduction that involves meiosis read **Extension 3.3** on alternation of generations. **X3.03S**

⬤ **Figure 3.24 A** Many organisms can regenerate parts of their body using mitosis. **A** The crown-of-thorns starfish eats coral and has caused considerable damage to some coral reefs including the Great Barrier Reef off Australia. Early attempts to control this starfish included chopping them up and throwing the bits back into the sea. Unfortunately, due to their ability to regenerate new individuals from each piece, this just increased their numbers. **B** Freshwater *Hydra* attached to water weed. Notice the daughter that has budded from the parent by asexual reproduction.

Early embryonic development – stem cells

Fertilisation is followed by cleavage, a series of rapid cell divisions. In these first few divisions, the embryo divides without growing in size. Smaller cells are produced with each successive division. It is the large reserves of nutrients and extra cell contents that human eggs accumulate while in the ovaries that allows the zygote to divide rapidly soon after fertilisation.

Q3.11 Which phases of the cell cycle will be shortened to achieve the rapid division during cleavage?

Cells in the early embryo

After a human zygote has undergone three complete cell cycles, it consists of eight identical cells. Each of these cells is said to be **totipotent** as it can develop into a complete, healthy human being. This is what happens when identical twins (or triplets) form – such twinning can occur up to 14 days after conception. (These cells are called 'totipotent' as they have the 'potential' to develop into a 'total' individual.)

By 5 days after conception, a hollow ball of cells called the **blastocyst** has formed (Figure 3.25). The outer blastocyst cell layer goes on to form the **placenta**. The inner cell mass, of 50 or so cells, goes on to form the tissues of the developing embryo. These 50 cells are known as **pluripotent embryonic stem cells**. Each of these cells can potentially give rise to most cell types, though they cannot each give rise to all the 216 different cell types that make up an adult human body. (The 'pluri' of 'pluripotent' is like the word 'plural'. It means that the cell can give rise to many cell types but, unlike the earlier totipotent cells, not all of them.)

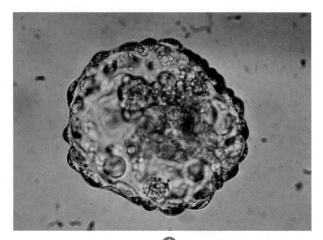

▲ **Figure 3.25** Five days or so after conception, a human zygote divides to form a hollow ball of cells. Between 6 and 12 days after fertilisation the embryo implants in the wall of the uterus.
Magnification ×215.

Cells become more differentiated

As the embryo develops, the cells of which it is made become increasingly **differentiated**. Most of them lose the capacity to develop into a wide range of cells. Instead they become increasingly specialised, functioning as a red blood cell, one of the cell types in bone, or whatever (Figure 3.29). However, even in adults some cells retain a certain capacity to give rise to a variety of different cell types. These cells are known as **multipotent** stem cells. For example, neural stem cells can develop into the various types of cell found in the nervous system, while blood stem cells, located in bone marrow, can develop into red blood cells, platelets and the various sorts of white blood cells (macrophages, lymphocytes and so on).

The presence of stem cells in an adult is not unique to humans, or even to mammals. For example, in the body wall of cnidarians, such as sea anemones and jellyfish, there are cells that on first sight appear to have no function. They do not help to defend the animal, feed it or protect it. They are just small, unspecialised cells.

Cnidarians have specialised cells that they use for paralysing their prey. These specialised cells work rather like harpoons and can only be fired once. They need to be constantly replaced and this is the role of the unspecialised cells, which divide and differentiate when more of the harpoon-like cells are needed.

Potential use of stem cells in medicine

Stem cells offer great hope to medicine. They may one day produce universal human donor cells which would provide new cells, tissues or organs for treatment and repair by transplantation. Embryonic stem cells may be the most suitable type of stem cells for this sort of treatment. Their potential to develop into any cell type offers the greatest flexibility for development, unlike adult stem cells which are committed to developing only into certain cell types.

Pluripotent stem cells for research and medicine would be isolated from so-called 'spare embryos'. These are produced in infertility clinics that carry out *in vitro* fertilisation, where the ovum is fertilised outside the body. Women undergoing this treatment are given drugs to make them superovulate, producing more eggs than are needed for infertility treatment. Most of the embryos end up being placed in women's wombs in the hope of enabling infertile couples to have children, but any additional embryos could be a source of embryonic stem cells. The use of embryos in this way is controversial – see the section in Topic 2 about ethical concerns (pages 93–95).

In the laboratory, embryos would be allowed to grow to form blastocysts. At this point, the embryos would be cultured for a further period of time to see if stem cells are formed. The stem cells would be isolated from each embryo and the rest of the embryo (which could not develop further) would be discarded. The stem cells would then be cultured and, hopefully, develop into tissues that could be used for transplantation (Figure 3.26).

Figure 3.26 Stem cells can be grown in culture with the potential to produce tissues or organs for transplantation. Technically, though, this is difficult and the work is still very much at the research stage. It is also controversial. This picture shows a culture flask of embryonic stem cells in a growth medium.

One problem with this approach is that even if the scientists manage to get the stem cells to develop into the right sort of tissue, that tissue might end up being rejected by the immune system of the person given the transplant.

There are several ways in which it is hoped to get round this problem of transplant rejection. One way would be to use tissue typing. This practice has routinely been used for decades when finding a suitable donor for a blood transfusion.

Did you know? Stem cells and therapeutic cloning

Tissue typing for the transplantation of whole organs is more complicated than for blood transfusions, but the principles are the same. If whole organs are one day grown from stem cells – and this is a very long way off – it has been estimated that perhaps only 20 stem cell lines would be needed to provide transplants for 90% of the UK population.

A second approach to the problem of rejection would be to use drugs that prevent the recipient from rejecting any transplanted organ. Again, this approach has been around for many years and is widely used with conventional organ transplants.

A third and novel approach has been proposed and is generally referred to as **therapeutic cloning**, though the use of this term is regretted by many of the scientists working in the area. In therapeutic cloning the patient needing a transplant would have one of their diploid cells removed – this could simply be a cell from the base of a hair or any other suitable tissue. This cell, or its nucleus, would then be fused with an ovum from which the haploid nucleus had

been removed (Figure 3.27). The result would be a diploid cell rather like a zygote.

This cell would then be stimulated to divide by mitosis in the same way as the cell which gave rise to Dolly the cloned sheep (see Figure 3.31, page 120). If all went to plan, after about 5 days a blastocyst would develop. Stem cells could then be isolated from this, and encouraged to develop into tissues. This procedure results in cell lines, and perhaps eventually organs for transplantation, which are genetically identical with the patient from whom the original diploid cell was taken.

As it is possible to maintain embryonic stem cells in tissue culture (Figure 3.26), it should also be possible to use them to find out how genes control human development. For example, they might be used to explore how genes trigger the onset of organ formation. They may help us to understand how cancer cells develop and how certain birth defects occur. Stem cells could also be used to provide a source of normal human cells of virtually any tissue type for use in screening new drugs.

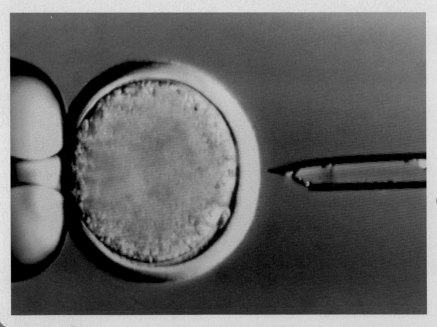

◀ **Figure 3.27** The ovum is held in place with a pipette. Its genetic material is removed and replaced with an adult cell nucleus. The first UK license to allow cloning of human embryos for stem cell research was granted in August 2004. Magnification ×565.

Ethical concerns about the use of stem cells

Just about everyone agrees that there are no ethical objections to using multipotent stem cells derived from adults, for example from adult bone marrow. The problem is that most scientists believe that these stem cells are likely to be less valuable for research and in developing new treatments than are the pluripotent stem cells which can only be derived from human embryos. Even if adult stem cells are used for new research, they may be of little use unless a better understanding is gained of how they specialise. This understanding may only come from embryonic stem cells.

Different people see the status of the human embryo very differently. As the official UK government committee set up to report on human stem cell research – with the Chief Medical Officer, Professor Liam Donaldson, in the chair – said in the year 2000:

A significant body of opinion holds that, as a moral principle, the use of any embryo for research purposes is unethical and unacceptable on the grounds that an embryo should be accorded full human status from the moment of its creation. At the other end of the spectrum, some argue that the embryo requires and deserves no particular moral attention whatsoever. Others accept the special status of an embryo as a potential human being, yet argue that the respect due to the embryo increases as it develops and that this respect, in the early stages in particular, may properly be weighed against the potential benefits arising from the proposed research.

In the UK, the Human Fertilisation and Embryology Authority (HFEA) regulates research on human embryos under the 1990 Human Fertilisation and Embryology Act. Until 2001, UK law only allowed the use of human embryos where the HFEA considered their use to be necessary or desirable:
- to promote advances in the treatment of infertility
- to increase knowledge about the causes of congenital disease
- to increase knowledge about the causes of miscarriage
- to develop more effective methods of contraception
- to develop methods for detecting gene or chromosome abnormalities in embryos before implantation.

On 22 January 2001, peers in the UK House of Lords voted by 212 to 92 votes to extend the purposes for which research on human embryos is allowed. (A vote in the House of Commons had already gone the same way in December 2000 by 366 to 174 votes.) So-called 'spare' embryos from *in vitro* fertilisation treatment can now also be used as a source of embryonic stem cells for the purpose of research into serious disease. On 27 February 2002 this was extended to include fundamental research necessary to understand differentiation and dedifferentiation of cells.

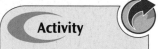

Activity

In **Activity 3.10** you can read some of the quotes from those who are for and against the use of stem cells and decide what your own position is on the issue. **A3.10S**

3.3 How is development controlled?

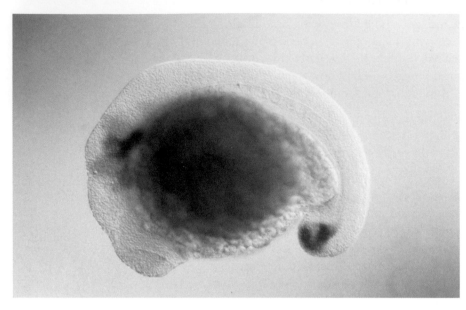

Figure 3.28 A zebrafish during an early stage in its development.

Extension

Using the weblinks in **Extension 3.4** you can watch the development of zebrafish and find out how these fish are being used to study development and genetics. You can also see the early stages in human development. **X3.04S**

In Figure 3.28 you can see an early stage in the development of a zebrafish. It is possible to watch the growth and development of these organisms through a light microscope because they are transparent, their eggs are fertilised externally and they develop very quickly in water.

As cells divide after fertilisation they become specialised for a small number of functions (Figure 3.29). As you probably already know, specialised cells often work together as tissues in organs.

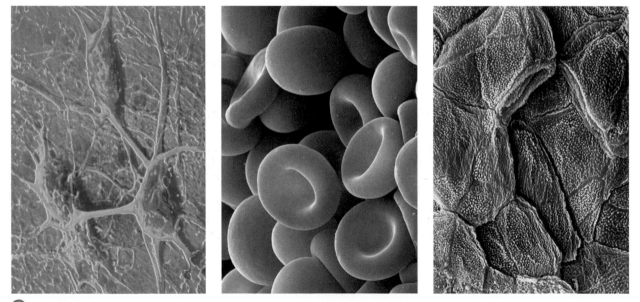

Figure 3.29 Although these specialised cells look different to one another, they have all been produced by mitosis from a common ancestor so we know they must all contain a complete complement of chromosomes with the same genetic information. Magnification L–R: ×4100, ×4600, ×2650.

We now take it for granted that the nucleus has a role in controlling the development of the individual cell and the whole multicellular organism's phenotype. This was first shown in classic experiments completed in 1943 by a Danish biologist named Joachim Hammerling (Figure 3.30).

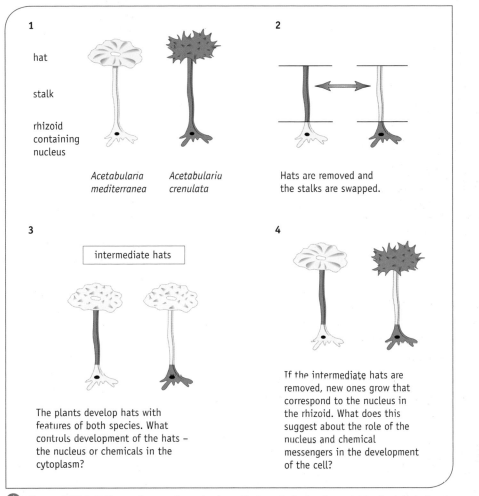

1

hat

stalk

rhizoid containing nucleus

Acetabularia mediterranea   *Acetabularia crenulata*

2

Hats are removed and the stalks are swapped.

3

intermediate hats

The plants develop hats with features of both species. What controls development of the hats – the nucleus or chemicals in the cytoplasm?

4

If the intermediate hats are removed, new ones grow that correspond to the nucleus in the rhizoid. What does this suggest about the role of the nucleus and chemical messengers in the development of the cell?

▲ **Figure 3.30** Is it the nucleus or the cytoplasm that controls development in *Acetabularia mediterranea* and *A. crenulata*?

Q3.12 In 1997, Ian Wilmut and his colleagues at the Roslin Institute in Scotland successfully cloned an adult sheep. They transplanted the nucleus from the mammary gland cell of one adult sheep into another sheep's unfertilised ovum from which the nucleus had been removed. The diploid cell that formed divided to form an embryo which was implanted into an adult sheep (Figure 3.31 on page 120). The surrogate mother gave birth to a lamb who soon became world famous, 'Dolly'.

How does the creation of Dolly support the idea that all the genetic information for making a complete organism is present in every single cell, including those in an adult with specialised functions?

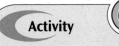

Activity

Using the simulation in **Activity 3.11** you can repeat Hammerling's experiments for yourself, demonstrating the role of the nucleus in controlling development of giant algal cells. **A3.11S**

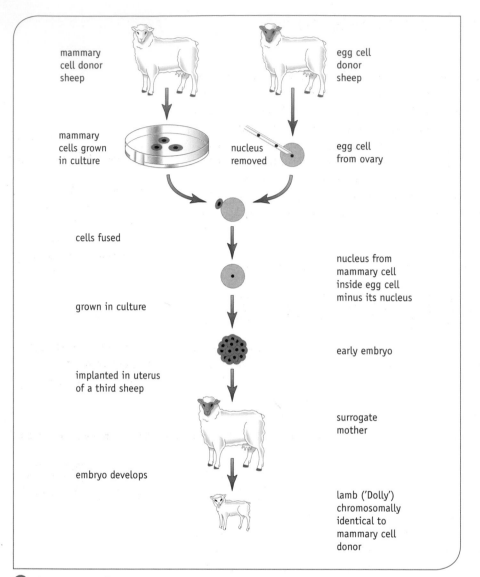

mammary
cell donor
sheep

egg cell
donor
sheep

mammary
cells grown
in culture

nucleus
removed

egg cell
from ovary

cells fused

nucleus from
mammary cell
inside egg cell
minus its nucleus

grown in culture

early embryo

implanted in uterus
of a third sheep

surrogate
mother

embryo develops

lamb ('Dolly')
chromosomally
identical to
mammary cell
donor

▲ **Figure 3.31** Dolly's genes were identical to those of the donor sheep.

Did you know? Problems with cloning

Animal cloning is not yet as easy as you might think from the above description. In fact most developmental biologists (including Ian Wilmut) feel that the process is not safe to try on humans. Many attempts to produce live-born animals by cloning have been unsuccessful and a high proportion of the animals produced by cloning have health problems. Dolly herself was the sole success from 277 attempts at cloning and this low success rate is still typical. Even Dolly may not have been completely 'normal' as she developed arthritis at a fairly young age for a sheep and had to be put down.

The reason why cloning mammals from adult nuclei is so inefficient is not yet fully understood, and there are probably several different contributing factors. One of these is that the DNA in an adult cell nucleus has been programmed into a particular type of cell (e.g. skin cell). When transferred to an ovum, the nucleus may not be able to reprogram its DNA fast enough to be able to switch on all the different genes required for normal development.

Reading the genome – which genes will be expressed?

As the embryo develops, cells differentiate: they become specialised for one function, or a group of functions. The function of each cell type is dependent on the proteins that it synthesises. This is determined by which genes in a cell are expressed, that is, transcribed and translated to produce the proteins they code for. For example, salivary gland cells make salivary amylase, an enzyme for digesting starch, while red blood cells contain the protein haemoglobin, an oxygen-carrying pigment. Amylase and haemoglobin are both proteins coded for by the cells' genes. How is it that these cells, and only these cells, express the genes coding for these proteins, although every cell contains the complete genome including the codes for both amylase and haemoglobin? What is it that switches on the genes appropriate for a cell's specialised function, and stops those that are not required from being expressed?

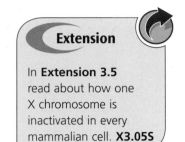

Extension

In **Extension 3.5** read about how one X chromosome is inactivated in every mammalian cell. **X3.05S**

Did you know? DNA inactivation by chemical modification

Chemical modification of the DNA can have a role in inactivation. Inactive genes have been found to have large numbers of methyl groups ($-CH_3$) attached to their DNA bases. If these methyl groups are removed, the genes become active. Once methyl groups have been attached to the DNA, they remain attached during subsequent divisions of the cell. This maintains the cell specialisation over successive mitotic divisions.

Key biological principle: What switches an individual gene on or off?

Genes in uncoiled, accessible regions of the DNA can be transcribed into messenger RNA. Transcription is initiated by an enzyme called RNA polymerase and a cluster of associated protein **transcription factors** binding to the DNA. They bind to a section of the DNA adjacent to the gene to be transcribed. This section is known as the promoter region. Only when the transcription initiation complex (RNA polymerase and transcription factors) has formed and been correctly attached to the DNA will transcription proceed. This process is outlined in Figure 3.32.

Some transcription factors are always present in all cells; others are synthesised only in a particular type of cell or at a particular stage of development. Many are created in an inactive form and are then converted into the active form by the action of signal proteins. The signal proteins may be hormones, growth factors or other regulatory molecules. The gene remains switched off until all the required transcription factors are present in their active forms. The transcription initiation complex can then form and attach to the promoter region successfully.

Transcription of a gene can be prevented by protein repressor molecules attaching to the DNA of the promoter region. This blocks the attachment sites for transcription factors, preventing binding of the transcription initiation complex. In addition, protein repressor molecules can attach to the transcription factors themselves, preventing them from forming the transcription initiation complex. In either case the gene is switched off; it is not transcribed within this cell.

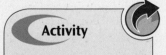

Activity

You can refresh your memory of protein synthesis by completing **Activity 3.12. A3.12S**

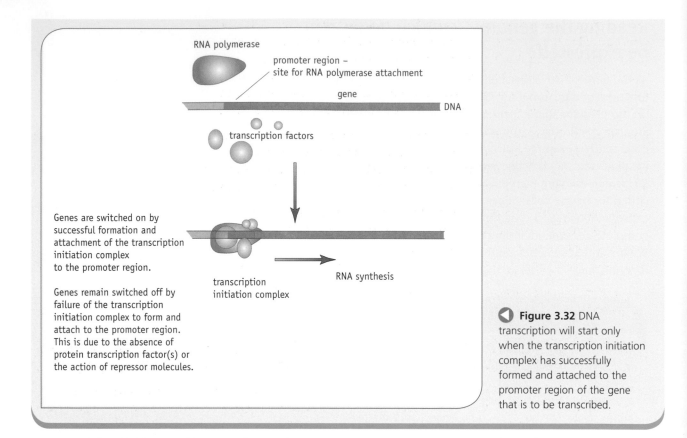

Genes are switched on by successful formation and attachment of the transcription initiation complex to the promoter region.

Genes remain switched off by failure of the transcription initiation complex to form and attach to the promoter region. This is due to the absence of protein transcription factor(s) or the action of repressor molecules.

Figure 3.32 DNA transcription will start only when the transcription initiation complex has successfully formed and attached to the promoter region of the gene that is to be transcribed.

How do signal proteins control growth and development?

Signal proteins may enter the cell and have a direct effect on the transcription factor (Figure 3.33A). More often, though, the protein attaches to the cell

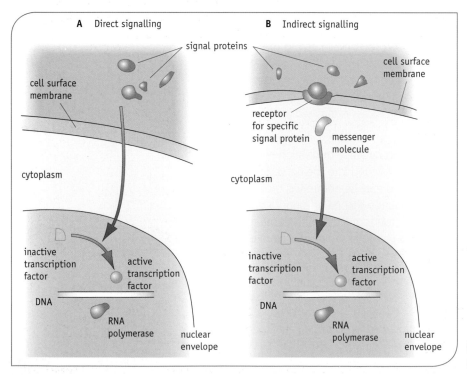

Figure 3.33 How signals affect transcription.

surface membrane and has an indirect effect, as shown in Figure 3.33B. The signal protein binds to receptors in the plasma membrane. Inside the cell, a second messenger molecule, which may be a protein, a lipid or a nucleotide, transmits the instruction through the cytoplasm and initiates transcription. Only cells with the appropriate receptors in the cell surface membrane and the necessary messenger molecules within the cell will respond to the signal.

It is important that signal proteins do not function inside the cells that produce them. For this reason, the protein is only activated once packaged into secretory vesicles by the Golgi apparatus. The vesicles fuse with the cell surface membrane, releasing their contents into the extracellular spaces by exocytosis.

Signal proteins are amongst the smallest proteins that occur in eukaryotic cells. For example, epidermal growth factor has only 53 amino acids. However, it has been shown that mouse epidermal growth factor is synthesised as a large protein precursor of 1168 amino acids.

Q3.13 Why is it necessary to produce an intercellular signal protein in an inactive form?

Key biological principle: How are genes regulated in prokaryotes?

The control of ß-galactosidase synthesis
The geneticists Jacob and Monod proposed a theory for the control of gene expression in the early 1960s. They studied the control of genes in the prokaryote *Escherichia coli*. These bacteria can use different carbohydrates as a food source, but they only produce the enzymes to break down a particular carbohydrate when that carbohydrate is present in the surrounding medium. To use lactose, the bacteria must produce the enzyme **ß-galactosidase**. This enzyme converts the disaccharide lactose to the monosaccharides glucose and galactose.

When lactose is not present in the environment, a lactose repressor molecule binds to the DNA and prevents the transcription of the β-galactosidase gene. Figure 3.34 shows how the lactose repressor stops the β-galactosidase gene being expressed. If lactose is present in the environment, the repressor molecule is prevented from binding to the DNA, and the β-galactosidase gene is expressed. In this case a biochemical signal in the form of lactose is switching on the gene. This is called induction.

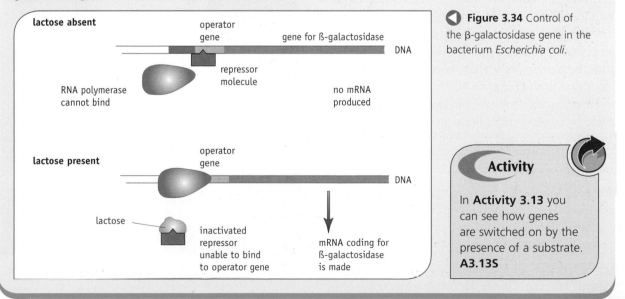

Figure 3.34 Control of the β-galactosidase gene in the bacterium *Escherichia coli*.

Activity

In **Activity 3.13** you can see how genes are switched on by the presence of a substrate. **A3.13S**

Did you know? How are cells organised into tissues?

Specialised cells can group themselves into clusters, working together as tissue. Cells have specific recognition proteins on their cell surface membranes, and these help them to recognise other cells like them and stick to them. A small part of each recognition protein is embedded in the cell surface membrane; a larger part extends from the membrane. This exposed section binds to similar proteins on the adjacent cell.

In the human embryo, interactions between cells to form tissues begin when cells start their specialised functions. At this point the genes coding for the recognition proteins are switched on.

Checkpoint

3.4 Write a short paragraph which summarises how transcription factors: **a** switch on and **b** switch off the transcription of a gene.

What happens when control of development goes wrong?

The importance of signals to development is very clearly illustrated when the signalling goes wrong.

Growing bones in the wrong place

Diana is a baby whose big toes are too short and are curved inwards towards her other toes. When her parents took her to a specialist they were told that this can be the first sign of a rare inherited disease called FOP (fibrodysplasia ossificans progressiva). FOP is characterised by the growth of bones in odd places, such as within muscle and connective tissue. In affected children, it begins with painful 'flare-ups' that accompany rapid abnormal bone growth. By early adulthood, this often leads to the 'freezing' of all major joints of the backbone and limbs so they cannot move.

Any local trauma, even injection of medicines into muscle, can cause bone formation in people with FOP (Figure 3.35). By their early 20s, most patients are confined to wheelchairs. Starvation used to result from the freezing of the jaw, and pneumonia can occur due to the lungs or diaphragm becoming fixed to the chest wall. Surgery can make the condition worse by causing more bone growth. There is currently no treatment.

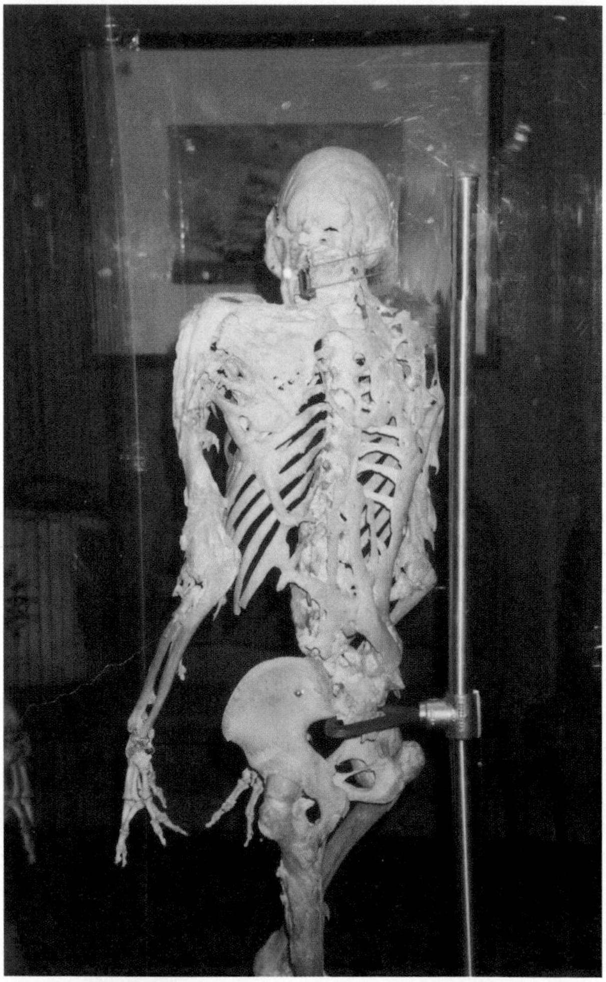

What causes FOP?

FOP is an inherited condition caused by a gene mutation. This results in the growth factor known as BMP-4 (bone morphogenic protein-4) being synthesised by the wrong cells.

BMP-4 acts as a signal protein and induces bone growth. The normal site for the synthesis of

▲ **Figure 3.35** Bones form in the wrong places in people with the rare inherited disease called FOP.

BMP-4 is in bone-producing stem cells of growing limbs and other places where the skeleton develops. In FOP, instead of being restricted to the places where bone should be found, BMP-4 is made by white blood cells. This is because a natural inhibitor of BMP-4 activity is missing. So BMP-4 is produced everywhere in the body where white blood cells travel, including inappropriate tissues such as muscle. In people with FOP, synthesis and release of BMP-4 results from tissue damage, as this activates the body's white blood cells.

The FOP story illustrates how throughout development both activating and inactivating mechanisms work together (like a see-saw). Inhibitors of BMP-4 are now being tested as a potential treatment for FOP. These inhibitors were originally identified as being important for amphibian neural development.

Q3.14 **a** Describe the role BMP-4 has in a cell of the skeleton.
b Describe how the 'natural inhibitor of BMP-4' might prevent transcription of the BMP-4 gene in cells that are not part of the skeleton.

How is the shape of an organism determined?

How do genes give instructions like 'this way is up' or 'now grow arms'?

Activities of the individual cells – division, differentiation, and in some instances death – all contribute to the 'big picture' of the organism's development, like pixels on a digital image. The transcription of genes in a set order determines the sequence of changes during development.

By studying the embryonic development of *Drosophila* fruit flies, the frog *Xenopus* and other model animals, researchers have shown how genes determine the sequence of changes during development.

It all starts in the egg – chemical gradients

Q3.15 Developmental biologists studying fruit flies, *Drosophila*, found there were a number of signal proteins in the embryo. Cells at one end of the embryo contained a high concentration of particular signal proteins, while the concentration of the same signal proteins was low in the cells towards the opposite end of the embryo. It is thought that these signal proteins originally enter the cell from the mother at one end of the embryo and diffuse across the cells of the embryo. The concentration of one such protein, signal protein A, is shown in Figure 3.36. Suggest what role signal protein A might have in determining the front of the embryo.

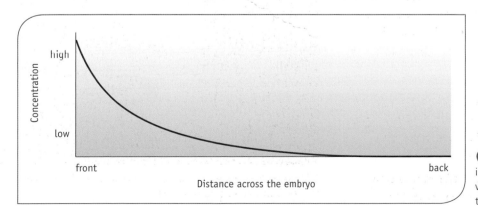

Figure 3.36 Protein gradients in the early embryo determine which end is the front, and which the back.

Q3.16 Once the main body segments have been determined, the appropriate structures develop in each segment. For example, in the case of *Drosophila*, antennae develop on the head, legs on the front, middle and back of the thorax, and wings at the back of the thorax. Master genes, called homeobox genes, control the differentiation of each segment. These genes were discovered by looking at mutants (Figure 3.37). All the proteins coded for by homeobox genes have a common sequence of 60 amino acids; this part of each protein binds to DNA.

a How might the proteins coded for by the homeobox genes cause the correct structures to develop in each body segment?

b Look at Figure 3.37. Describe in as much detail as you can what might have happened at the DNA level to the relevant homeobox gene to cause legs to be produced instead of antennae.

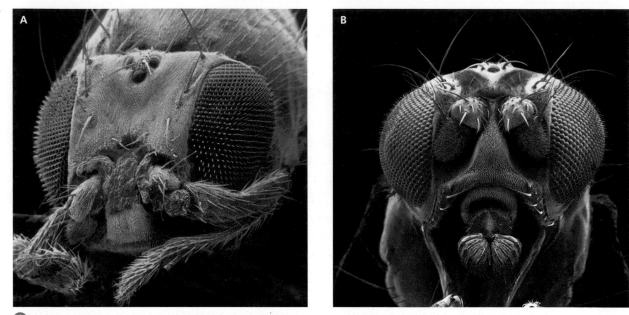

▲ **Figure 3.37 A** In this mutant fruit fly, legs have developed where the antennae should be. Magnification ×90. **B** The head of a normal fruit fly. Magnification ×115.

Did you know? Programmed cell death (apoptosis)

It is an amazing fact that all cells in the human body have a 'self-destruct' programme. During development an animal must lose some cells by programmed cell death, called apoptosis. There is a small group of cell 'suicide' genes, and when they are expressed this causes the nucleus and cytoplasm to fragment.

Genes that prevent death are expressed in most cells, and during development cells have to switch on their 'suicide' genes in order to die. This explains why cell death does not happen throughout the embryo. Programmed cell death occurs in particular places, such as between where the fingers will be. In the embryo brain many millions of extra cells are produced, at least half of which die by controlled cell death, leaving behind the precise pattern of adult connections.

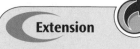

Extension

In **Extension 3.6** read how chemical gradients are created in embryos. **X3.06S**

3.4 Genes and environment

Is it all in the genes? Nature and nurture

The characteristics of an organism, such as its height, shape, blood group or sex, are known as its phenotype. Differences in phenotype between the members of a population are caused by differences in:

- genetic make-up or genotype
- the environment in which an individual develops.

Some characteristics are controlled almost completely by the organism's genotype, with the environment having little or no effect. For example, a person's blood group (group A, B, AB or O) is controlled by the genes they inherit. The genes code for protein antigens on the surface of their red blood cells. A person's blood group is not affected by the environment in which they grow up. Such characteristics show **discontinuous variation**. They have phenotypes that fall into discrete groups with no overlap (Figure 3.38A). In garden peas, unlike humans, height also shows discontinuous variation.

Characteristics that are affected by both genotype and environment often show **continuous variation**. Human height is a characteristic showing continuous variation. This means that a person can be any height within the human range. The most common height will be somewhere between the extremes of the range. We can show this pattern by asking a group of people to line up behind markers showing height categories, say from 1.5 m to 2 m in 2 cm steps. If the sample is big enough, the formation of people will reveal a bell shape when seen from above. Alternatively, a graph can be drawn showing the frequency distribution of the different height categories (Figure 3.38B).

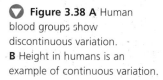 **Figure 3.38 A** Human blood groups show discontinuous variation. **B** Height in humans is an example of continuous variation.

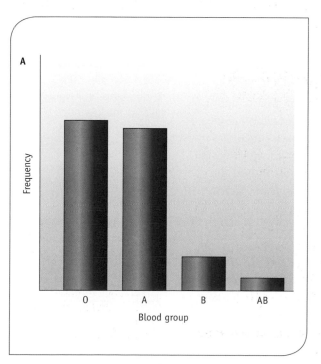

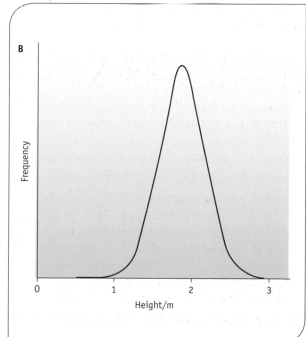

There are countless examples of genes and environment interacting to produce an organism's phenotype. Some are very familiar, such as skin and hair colour; others are less so. We will take a detailed look at three examples:

- height
- skin and hair colour
- causes of cancer.

Height

Have you noticed that each human generation seems to be a bit taller than the last (Figure 3.39)? The average human height in industrialised countries has risen each generation over the past 150 years. People in the UK now average about 8 cm taller than they did in 1850. How can this be explained?

Before reading on, list as many reasons as you can for the increase in human height over recent generations.

The following are some possible answers. You may have thought of others.

1 There is some evidence that taller men have more children, which would result in a gradual change in the genetic make-up of the population.

2 Greater movements of people have resulted in less inbreeding, which increases hybrid vigour (see Topic 4, page 174) leading to taller people.

3 Better nutrition, especially increased protein, has resulted in greater growth of children.

4 Improved health, with a reduction in infectious diseases, has occurred through improved sanitation, clean water supplies, vaccination and antibiotics.

5 The end of child labour has allowed more energy to go into growth.

6 Better heating of houses and better quality clothing reduces the amount of energy needed to heat the body, so again more energy can go into growth.

Q3.17 Which of these reasons for change in height are due to the environment and which are the result of genotype?

It is widely accepted that a person's height is determined by an interaction between the effects of their genes for height and environmental influences, such as diet. A person may have genes for being tall, but not achieve his or her full potential height because of malnutrition. We do not know for certain which of the possible reasons for change in height is the most important. However, most people think that better diet is the most significant factor.

Skin and hair colour

It can be dangerous to get burnt in the sun, but many people enjoy the feeling of the sun on the skin, especially after a long winter. Exposing skin to sunlight also helps to make vitamin D, which is needed for the uptake of calcium into teeth and bones. But why does sunlight make the skin darker?

▲ **Figure 3.39** People are getting taller.

Activity

Take part in **Activity 3.14** to see if the difference in heights over the years can be observed. **A3.14S**

Extension

In **Extension 3.7** find out how both genotype and environment influence sex determination. **X3.07S**

Making melanin

The dark pigment in skin is called **melanin**. To make melanin, animals use an enzyme called **tyrosinase**. Tyrosinase catalyses the first step along a chemical pathway, changing the amino acid tyrosine into melanin.

Melanin is made in special skin cells called melanocytes, which are activated by **melanocyte-stimulating hormone** (**MSH**). There are receptors for MSH on the surface of the melanocyte cells. Ultraviolet (UV) light increases the amount of MSH and also of MSH receptors, making the melanocytes more active (Figure 3.40). The melanocytes place melanin into organelles called melanosomes. The melanosomes are transferred to nearby skin cells where they collect around the nucleus, protecting their DNA from harmful UV light.

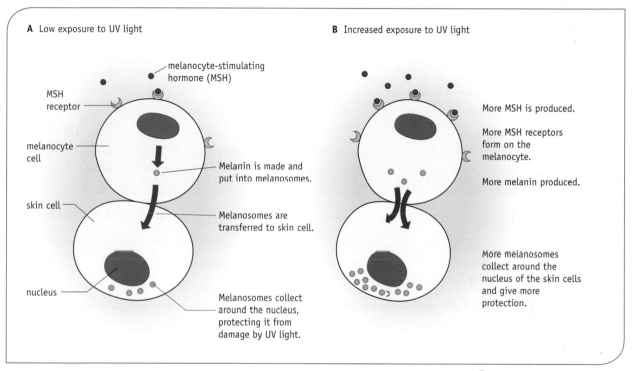

A Low exposure to UV light

melanocyte-stimulating hormone (MSH)

MSH receptor

melanocyte cell

Melanin is made and put into melanosomes.

skin cell

Melanosomes are transferred to skin cell.

nucleus

Melanosomes collect around the nucleus, protecting it from damage by UV light.

B Increased exposure to UV light

More MSH is produced.

More MSH receptors form on the melanocyte.

More melanin produced.

More melanosomes collect around the nucleus of the skin cells and give more protection.

▲ **Figure 3.40** The role of melanin in protecting the nucleus from UV light.

The number of MSH receptors in skin cells varies between different human races. People with more receptors have darker skins, giving more protection against sunburn.

Always white – albinos

Human albinos have a gene mutation which prevents them from producing melanin. They have white hair, white skin and no pigment in their eyes (iris and retina), making their vision poor. The phenotype for skin coloration in these individuals is determined solely by their genotype; the environment has no effect.

Q3.18 Many human albinos have two copies of a mutant allele for the tyrosinase gene, which codes for the enzyme involved in melanin production. Without tyrosinase no melanin can be made. There are several other genes which if mutated will also result in a lack of melanin. Suggest the role of these genes in melanin production.

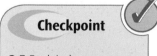

Checkpoint

3.5 Explain how a person's genotype and their environment can both affect the colour of their skin.

White with dark tips

Some animals, such as Himalayan rabbits and Siamese cats, have mutant alleles for tyrosinase. The enzyme is made but it is unstable and is inactivated at normal body temperature. However, the tips of their tail, paws, nose and ears are much darker than the rest of their bodies (Figure 3.41).

Q3.19 Explain how the environment and genotype are interacting to produce the distinct colouring at the tail, paw and ear tips in Figure 3.41.

Q3.20 There are rare cases in humans in which hair in the armpits is white, but hair on places such as the legs is dark. Suggest how this happens.

▲ **Figure 3.41** Only the ears, nose, paws and tail of this Himalayan rabbit are dark.

Seasonal colour change

Arctic foxes have brown fur in summer and white fur in winter (Figure 3.42). They must have the genes for making brown fur (which contains melanin) all the time, so how can white fur be made?

The white winter coat is actually grown during the summer. It grows under the brown summer coat and is revealed when the summer coat is moulted in autumn. The foxes produce fewer MSH receptors in the summer.

Q3.21 Explain why having very few MSH receptors means that the coat will grow white.

Q3.22 Explain why is it surprising that white fur is able to grow in the summer.

▼ **Figure 3.42** How does the arctic fox change its coat colour?

Cancer

What causes cancer?

One in three people in the UK will suffer from cancer at some stage in their life and, at present, one in four people dies from the disease. If we could understand the way our genetic make-up and our environment combine to cause cancer, we would be on the way to finding means of prevention and cure.

Cancers occur when the rate of cell multiplication is faster than the rate of cell death. This causes the growth of a **tumour**, often in tissues with a high rate of mitosis, such as the lung, bowel, gut or bone marrow. Cancers are usually caused by damage to DNA. DNA is easily damaged by physical factors such as UV light or asbestos. It can also be damaged by chemicals, known as carcinogens, which may be in the environment or can be produced by cell metabolism. If DNA is copied incorrectly in gamete formation, an inherited form of cancer can result.

Q3.23 Explain why cancers are more likely to occur in tissues with a high rate of mitosis.

Q3.24 Explain why damage to the DNA in an embryo can result in inherited cancer.

Telling cells what to do

Cells go through a fixed sequence of events during the cell cycle – G1, S, G2 and mitosis (M) – as shown on page 107. The progression from one phase to the next is controlled.

Cancer cells do not respond to the control mechanisms. Two types of gene have a role in control of the cell cycle and play a part in triggering cancer. These are:

- oncogenes
- tumour suppressor genes.

Oncogenes code for the proteins that stimulate the cell cycle. Mutations in these genes can lead to the cell cycle being continually active. This may cause excessive cell division, resulting in a tumour.

Tumour suppressor genes produce suppressor proteins that stop the cycle. Mutations inactivating these genes mean there is no brake on the cell cycle. One example of a tumour suppressor protein is p53. This protein stops the cell cycle by inhibiting the enzymes at the G1/S transition, preventing the cell from copying its DNA. In cancer cells a lack of p53 means the cell cannot stop entry into the S phase. Such cells have lost the control of the cell cycle. Loss of tumour suppressor proteins has been linked to skin, colon, bladder and breast cancers.

There is a very complex network of signals and inhibitors that interact to control the cell cycle. There needs to be damage to more than one part of the cell control system for cancer to occur. This makes cancer very unlikely in any particular cell, but because the body contains so many cells dividing and changing over a lifetime, cancers will occasionally occur. Cancers are more likely in older people as they have accumulated more mutations.

Did you know? How enzymes control the cell cycle at checkpoints

Progress through the cell cycle is controlled at checkpoints between the stages G1, S, G2 and M. At each checkpoint the formation of proteins called cyclins controls the passage to the next stage. The cyclins activate enzymes called cyclin-dependent kinases (CDKs). These enzymes initiate reactions that occur in the next phase of the cycle.

Figure 3.43 shows the role of CDK1 in initiating prophase. Cyclins accumulate during G2 of interphase. They attach to the CDK and the complex formed catalyses phosphorylation of other proteins. Phosphate added to the proteins changes their shape and makes them active. For example, phosphorylation of proteins associated with our DNA can lead to condensation of the chromosomes.

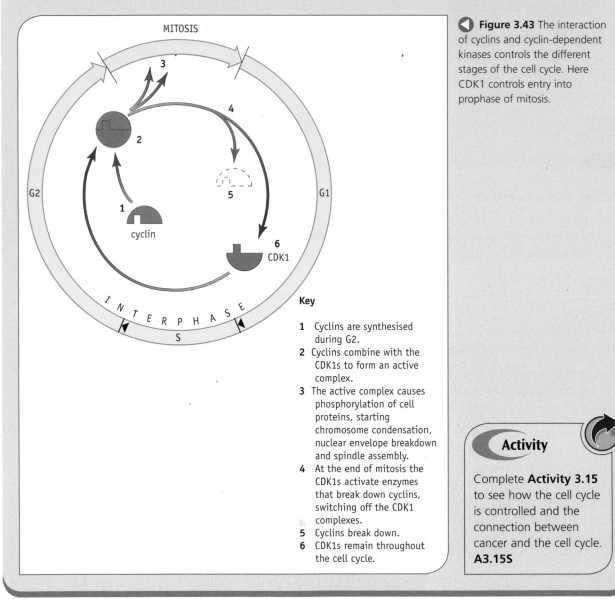

◀ **Figure 3.43** The interaction of cyclins and cyclin-dependent kinases controls the different stages of the cell cycle. Here CDK1 controls entry into prophase of mitosis.

Key

1 Cyclins are synthesised during G2.
2 Cyclins combine with the CDK1s to form an active complex.
3 The active complex causes phosphorylation of cell proteins, starting chromosome condensation, nuclear envelope breakdown and spindle assembly.
4 At the end of mitosis the CDK1s activate enzymes that break down cyclins, switching off the CDK1 complexes.
5 Cyclins break down.
6 CDK1s remain throughout the cell cycle.

Activity

Complete **Activity 3.15** to see how the cell cycle is controlled and the connection between cancer and the cell cycle. **A3.15S**

Q3.25 In some people one of the two alleles for the tumour suppressor protein p53 is damaged. The damaged allele is recessive to the normal allele. Explain why such people are more susceptible to environmentally induced cancer than people with two normal alleles for p53.

Inherited cancer

About 5% of cancers occur because of an inherited gene. For example, mutations in the gene BRCA1 predispose a person to breast cancer. The functioning BRCA1 gene produces a protein used to repair DNA. A child born with one defective BRCA1 allele may get cancer later in life if the other allele becomes damaged in breast tissue cells. Having a single defective BRCA1 allele does not therefore mean that breast cancer is inevitable. It simply means that such individuals are more susceptible to cancer through environmental DNA damage. There are other genes which predispose people to many other cancers including bowel cancer, prostate cancer, retinal cancer and some types of leukaemia.

Environmental causes of cancer

You will have noticed that newspapers and magazines are full of suggestions for living a healthy life, and reducing cancer risk. How useful are these suggestions? Damage from the environment can be either chemical or physical. The greatest chemical risk of all is from smoking, which increases the likelihood of many forms of cancer, especially lung cancer, through the action of carcinogens in tar. Tar lodges in the bronchi and causes damage to DNA in the surrounding epithelial cells.

Ultraviolet light can physically damage DNA in skin cells. A mole may start to grow bigger, and can develop into a tumour (Figure 3.44). If a tumour is not removed, cancer cells can sometimes spread to other parts of the body carried in the blood and lymphatic systems, causing new cancers in other organs. This spreading of cancer is known as metastasis.

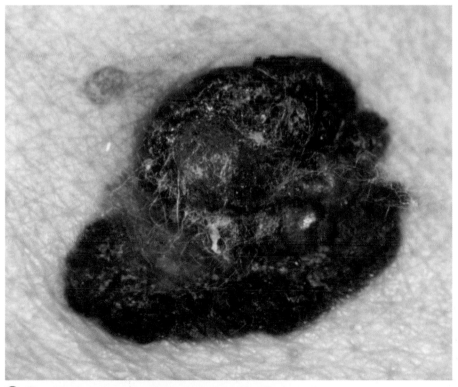

Figure 3.44 UV damage can cause cancer in skin cells.

Diet is also linked to cause and prevention of cancer, though the connections are not always clear. A diet with plenty of fresh fruit and vegetables provides antioxidants which destroy radicals (see Topic 1). Radicals are chemicals produced by the cell's own metabolism which contribute to ageing and cancer through DNA damage.

Several cancers are triggered by virus infection. For example, liver cancer can follow some types of hepatitis, and cervical cancer can follow infection by the papilloma (genital wart) virus. A virus's RNA may even contain an oncogene, which it has picked up from one of its hosts and then transfers to the cells it infects.

Did you know? Combating cancer

One way of treating cancer is to use surgery. Another is to destroy the cells in the tumours. In chemotherapy (Figure 3.45) powerful chemicals are used to do this, and in radiotherapy, X-rays or other radiation are directed at the tumour. The difficulty is to target the tumour cells without damaging nearby healthy tissues.

It used to be thought that chemotherapy and radiotherapy work by preferentially killing dividing cells in the tumour. It is now believed that these treatments actually work by inducing cells to carry out cell suicide. Chemotherapy and radiotherapy cause some DNA damage, but not enough to kill tumour cells. However, the DNA damage causes the release of p53, so the cancer cells stop dividing.

Q3.26 Explain why chemotherapy and radiotherapy are often unsuccessful in tumours where the cause of the cancer is damage to the protein p53.

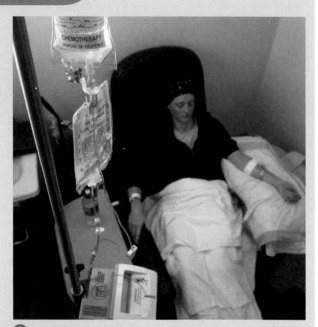

▲ **Figure 3.45** A wide range of chemotheropy drugs is available. Here, a patient is receiving anti-cancer drugs intravenously via a drip.

3.5 Will knowing our genetic code help?

The recent deciphering of the base sequence in the human genome as part of the Human Genome Project means that we are likely to gain a much better understanding of the way genes control our phenotype. An optimistic prediction is that we will then be in a better position to avoid the risk factors specific to our own genetic make-up, and to improve the help given to people with problems that have a genetic basis.

The Human Genome Project

A genome is all the DNA of an organism (or species), including the genes that carry all the information for making the array of proteins required by the organism (or species). It is these proteins that help determine all the characteristics of the organism, from the individual biochemical pathways to its overall appearance.

Finding the sequence of bases in DNA

In 1977, Fred Sanger invented the first DNA sequencing process. In this process DNA is used as a template to replicate a set of DNA fragments, each differing in length by one base. The fragments are separated according to size using gel electrophoresis and the base at the end of each fragment is identified. This allows the sequence of bases in the whole DNA chain to be determined.

Following this it became possible to work out the entire sequence of bases in the human genome, and determine the location of all our genes. This was a massive undertaking, as Figure 3.46 illustrates.

Figure 3.46 The human genome contains a total of some three thousand million (three billion, i.e. 3×10^9) bases. It's hard to imagine such a huge number, so try this: 15 four-drawer filing cabinets, each with drawers full of A4 paper, printed on both sides with the letters ACGTTGTACAGTG ... typed in 12 point.

In 1986 the Human Genome Project was officially started, with the USA and UK being full partners. It was thought that the task of deciphering the whole of the human genome would take over 20 years, but with the rest of Europe and Japan joining the project in 1992, and the development of sophisticated computerised systems (Figure 3.47), progress was more rapid.

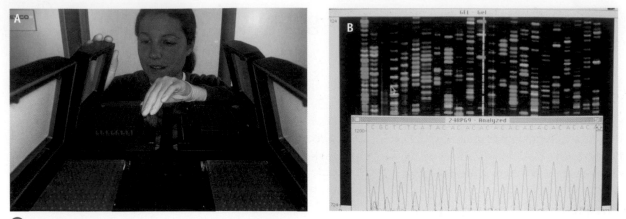

▲ **Figure 3.47 A** DNA sequencing is highly automated, with computers collating the sequence. **B** DNA sequence on a computer screen. Each of the four bases is shown by a different colour, the order of colours giving the base sequence.

Despite its name the project did not sequence only the human genome. The complete DNA sequence of yeast was published in 1996 and this was followed in 1998 by that of the nematode worm *Caenorhabditis elegans*. Other organisms used in biological and medical research, such as *Drosophila*, zebrafish, the mouse and the rat have also been sequenced.

In 1999 chromosome number 22 became the first human chromosome to be fully sequenced. A working draft of the whole human genome was published in 2001. The draft contained gaps and had an error rate of about 1 in 1000 bases. Work continues with sections being sequenced up to ten times to produce a high-quality finished sequence.

> **Extension**
>
> In **Extension 3.8** and the associated weblinks you can find out more about DNA sequencing and the Human Genome Project. **X3.08S**

The base sequence is just the beginning ...

The aim of the Human Genome Project (HGP) has been to produce a high-quality sequence of the bases in our DNA without gaps and with 99.99% accuracy. However, the sequence itself is really only the starting point; it provides a resource for biologists across the world. The policy of the publicly funded project was to immediately publish the sequence on the Internet making it freely available to everyone. As a result analysis and interpretation of the sequence could start even before the final sequence was published.

Using the sequence biologists are gaining a better understanding of the genome itself, identifying new genes, working out how they are controlled, and discovering what products they code for. The sequence also has a major part to play in research into the role of genes in disease and the development of new diagnostic techniques and treatments. What we learn about the DNA in other organisms will open up possibilities of solving other challenges in health care, finding improved agricultural methods and cleaning up environmental pollution. The potential is staggering, but we are only just at the beginning.

Outcomes of the Human Genome Project

Detailed information about the genome

The human genome is now thought to be 3 200 000 000 bases in length. It is estimated that our genome contains 30 000 to 40 000 genes, which are more complex than those of simpler organisms and give rise to a large number of more complex proteins.

The average human gene is now known to consist of about 3000 bases, but sizes vary greatly. The largest known human gene codes for dystrophin; it is 2.4 million bases long. The dystrophin protein is faulty in individuals with the muscle-wasting disease Duchenne muscular dystrophy. However, the functions of most of our genes are still unknown.

Non-coding sequences ('junk DNA') make up at least 50% of the human genome. Once considered to have no direct function, they are now thought to be important in gene regulation. Scientists have identified about 1.4 million locations where single-nucleotide polymorphisms (SNPs – pronounced 'snips') occur in humans. A SNP is a DNA sequence variation that involves a change in a single nucleotide. SNPs are used to help locate disease-associated DNA.

Identification of new genes

Many disease genes have been identified, including the breast cancer gene BCRA2 and the total colour blindness genes CNGA3 and CNGB3 (these code for subunits of a protein channel in photoreceptor cells in the retina). It is now possible to locate a candidate gene on our DNA (a gene that may cause a particular disease) and then screen this gene for mutations in affected individuals. Analysing DNA sequence patterns in humans side by side with those in well studied model organisms (such as yeast or *Drosophila*) has become one of the most powerful strategies for identifying human genes and interpreting their functions.

Identification of new drug targets

A drug target is a specific molecule that a drug interacts with to bring about its effect. Before the entire draft human DNA sequence was published, there were only 500 drug targets. The identification of disease genes and their products should allow biologists to find new drug targets. By the time the draft sequence was published in 2000, scientists had already searched for DNA sequences similar to those for existing drug target proteins, and found 18 new sequences that may potentially provide new targets.

Preventative medicine and improved drug treatment

Some drugs work very well for one person with few side effects, whereas in other patients the same drug is ineffective and may have major side effects. It is thought that these different responses may be due to slight variations in each individual's genome, according to which of the 1.4 million SNPs a person possesses. Prescribing the best drug for a patient is currently largely trial and error. It is hoped that information about a person's genome will enable doctors to prescribe the right drug at the correct dose.

If a person knew that they carried mutations associated with a particular disease, they could make changes to their lifestyle that would reduce the risk of the disease, or opt for preventative treatment. For example, if we knew the gene mutations associated with asthma and we knew the environmental factors that trigger attacks, children with the mutations might be able to avoid exposure to the environmental factor. Alternatively they might be exposed in a controlled way in their early years which could perhaps avoid triggering a lifelong allergy.

Understanding basic biology

In addition to medical applications, the DNA sequence contributes to research in many areas of cell biology and physiology. For example, researchers investigating the molecular basis for why some people cannot detect bitter tastes used the DNA sequence which was freely available. By identifying the sections of DNA involved and searching the DNA sequence, they discovered receptor proteins that are found in taste buds. Experiments using cultured cells confirmed that the receptors do respond to the bitter taste.

Studying proteins and their functions – known as proteonomics – can bring researchers closer to what's actually happening in the cell. Recent work has shown that single genes can produce a range of different protein products as a result of varying post-production processing, and that most proteins work together in groups. It is now being recognised that the total number of proteins found in the cell, the proteome, is far greater than the number of genes in the genome.

Investigating evolution of the human genome

A surprisingly large proportion of our DNA consists of repeat sequences. These are thought to have no direct function, but shed light on chromosome structure and the history of our DNA. Over time, these repeats replicate and insert new copies of themselves at different locations in the genome. They reshape the genome by rearranging it, creating entirely new genes, and modifying and reshuffling existing genes.

By comparing our genome with those of other animals it is possible to look at the evolutionary history of the genome, determining when particular genes first appeared and seeing how mutations accumulate over time. The same genes are almost always found in other vertebrate species, strongly suggesting a common ancestor in evolution. The occurrence of SNPs will also provide a useful tool for investigating human evolution.

Ethical dilemmas

The human genome sequence offers great hope for the future, but there is also considerable concern about the potential use of this information. In 1991, shortly after the Human Genome Project got underway, the Director of the National Institute of Health (NIH) in America filed applications for patents on 6122 fragments of DNA. As a result, James Watson resigned from his position as leader of the American branch of the Human Genome Project. The President of the USA and the UK Prime Minister issued a statement in March 2000, declaring that all genetic data should be released into the public domain and not restricted by patent. However, patents on human DNA are still being taken out.

One of the main project goals of the Human Genome Project is to address the ethical, legal and social issues that may arise from the project. A proportion of the budget (3–5%) has been set aside for this purpose. However, many issues remain to be addressed adequately:

- Testing for genetic predisposition has many implications. Would it be acceptable, for example, for insurers to have this information about people who are applying for health insurance?
- Who should decide about the use of genetic predisposition tests, and on whom they should be used?
- Making and keeping records of individual genotypes raises acute problems of confidentiality.
- Through the developing genetic technologies, many medical treatments made possible will initially be very expensive. Their restricted availability will add considerably to the problems faced by the health services in deciding who is eligible for such treatments.

The breast cancer story

Breast cancer (Figure 3.48) is the most common cancer in women, accounting for 30% of all women's cancers. (Only 1% of breast cancer cases occur in men, with about 300 men being diagnosed with the disease each year, most of them over 60 years old.) For women in the UK there is a 1 in 9 lifetime risk of developing breast cancer. Although the number of deaths has been falling steadily in recent years there are still over 10 000 women who die from the disease each year in the UK.

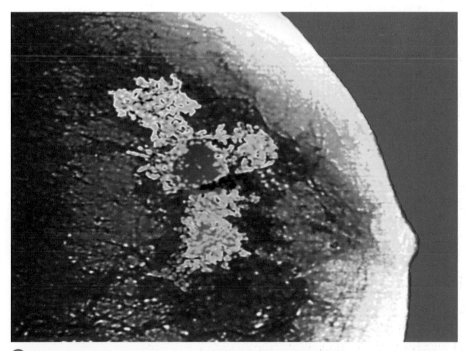

▲ **Figure 3.48** Coloured mammogram (X-ray) showing cancerous tumour (blue). It is worth noting that nine out of ten breast lumps are not a sign of breast cancer but any that appear should be checked by a doctor.

We have seen that mutations in the BRCA1 gene are linked to a higher risk of developing breast cancer at an early age. The human genome sequence was used when identifying the second breast cancer gene, BRCA2. First the possible location of the BRCA2 gene was identified by tracking known DNA markers through members of families affected by the disease. Then segments of this section of DNA were cloned and sequenced. The human genome sequence data was used to confirm its position on chromosome 13.

Several mutations in the gene were detected by analysing DNA from breast cancer patients. These mutations all cause premature termination codons. These result in shortened proteins which are unable to complete their function in the repair of DNA.

Will I get breast cancer?

The singer Linda McCartney died from breast cancer, whereas Olivia Newton-John, star of the film *Grease*, survived the disease. Their daughters, like the thousands of other daughters whose mothers have had breast cancer, face the question – will I get the disease?

Should women whose mothers have had breast cancer have a test for the gene mutations associated with the disease?

Unfortunately it is not as simple as this. Only 5% of breast cancer is hereditary; 95% of cases show no genetic link. These cases are called sporadic. So even if tested and found not to have any of the 235 mutations in the BRCA1 gene and 100 mutations in the BRCA2 gene, there is no guarantee that the individual will not get breast cancer. It is also the case that a third of inherited cases show no mutations in the BRCA genes, suggesting that other genes may be involved.

However, genetic testing may be an option to consider in a high-risk family where several relatives have been affected by breast and/or ovarian cancer at an early age. Females in such families have an 80% chance of developing the disease by age 70. The most reliable method of testing is to sequence whole genes and look for mutations, but this may be expensive and time consuming. If particular mutations can be identified in affected family members, testing relatives for these specific mutations is easier and cheaper. Finding that they did not have the mutations would reduce anxiety for these relatives. However, they could still get breast cancer as a result of other mutations or other causes. Alternatively, testing for the gene mutations and finding one of them present would allow the individual to make decisions about possible courses of action. They may opt for regular breast examination and mammograms (breast X-rays) to identify any early signs of the disease, thus ensuring early treatment which has a higher chance of success. Or they may choose preventative treatment such as surgery which can greatly reduce though not entirely eliminate the risk.

Genetic testing and discrimination

Women (or men) identified as carriers of the genes associated with breast cancer or other inherited diseases may face discrimination in employment or when applying for insurance. Individuals may be faced with higher premiums or be unable to get certain types of cover even though the possession of the gene does not mean that the person will necessarily develop the condition. This is particularly true in the US where the majority of medical treatment is paid for through health insurance. In 2003 the US Senate unanimously voted for the Genetic Information Nondiscrimination Act 2003. This Act prohibits the use of genetic information by insurers or employers, but must also be passed by the House of Representatives before it becomes law. In October 2001 the UK Government announced a five-year moratorium on the use of genetic test results in assessing applications for some insurance policies. However, at the time of writing (2005) there is no Act of Parliament prohibiting this type of discrimination.

Q3.27 Suggest reasons why carriers of breast cancer genes should not face discrimination when applying for jobs or insurance.

People may decide against taking a genetic test in case it results in discrimination against them. They may decide to avoid anxiety by not being tested, exercising their right 'not to know'. However, other family members might argue that if there are implications for them they also have a right to know. Genetic counselling helps individuals make decisions about genetic testing (Figure 3.49) and some people decide not to have genetic testing as a result of such counselling.

Figure 3.49 Counselling can help people make informed decisions about genetic testing.

Genetic testing and prenatal diagnosis

As we saw in Topic 2, prenatal diagnosis allows parents to find out if a fetus is carrying a serious inherited condition. When an embryo has developed into a ball of undifferentiated cells, it is possible to remove a single cell for genetic screening without harming the embryo. This technique has been used to select for female embryos in families with a history of a **sex-linked** inherited disease, such as Duchenne muscular dystrophy (Figure 3.50). Such diseases are much more likely to be expressed in males, so male embryos are discarded – even though half of them would be born unaffected by the disease. Genetic screening has also been used to produce a child who is a suitable bone marrow donor for an existing child with a medical disorder.

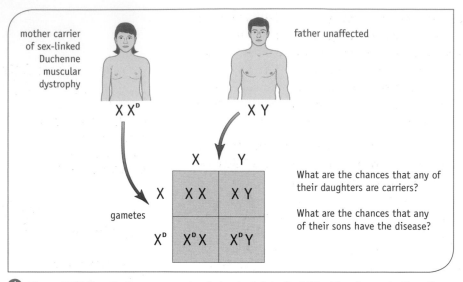

mother carrier of sex-linked Duchenne muscular dystrophy

X Xᴰ

father unaffected

X Y

gametes

	X	Y
X	X X	X Y
Xᴰ	Xᴰ X	Xᴰ Y

What are the chances that any of their daughters are carriers?

What are the chances that any of their sons have the disease?

△ **Figure 3.50** How Duchenne muscular dystrophy is inherited. Would embryo selection allow families to avoid passing on the disease?

More widespread testing for gene mutations raises concerns about who determines what is a serious condition, and the selection of characteristics by parents to create 'designer babies'.

It is not yet, and may never be, possible to select an embryo with a high potential IQ, since so many different genes are involved in intelligence. As we have seen, the environment is usually involved in determining phenotypic characteristics that show continuous variation, such as IQ and height. So simply selecting embryos with the genes for high IQ may not be enough, even if we could identify them. However, it is becoming increasingly feasible to select embryos for phenotypic characteristics that are inherited in a simple way, such as sex or eye colour, and to discard embryos with unwanted alleles, such as those for Huntington's disease or cystic fibrosis.

Checkpoint

3.6 List the key social and ethical issues raised by the use of genetic screening in humans.

Eugenics

The genetic selection of humans is known as **eugenics**. Eugenics was widely advocated in many Western countries in the first half of the twentieth century as a means of 'improving the genetic stock' of the human race, but fell into disrepute following the persecution and elimination of minorities in Nazi Germany. However, eugenics by individual choice occurs today in at least three contexts:

● selectively aborting embryos after genetic screening in the womb
● embryo selection in IVF
● providing sperm or eggs to infertile couples from a chosen donor.

Q3.28 Explain why each of the three practices listed above is an example of eugenics.

Eugenics could be taken a stage further by inserting desired genes into the embryos used in IVF treatment. This is **germ-line gene therapy**. It is not permitted in humans and would currently be irresponsible to attempt because of the risks involved, since any genetic problems arising could be passed on to future generations. However, such techniques are already widely used in non-human animals. For example, **transgenic** pigs have been created which may eventually be used as a source of donor hearts for human transplants.

Eugenics in general and germ-line gene therapy in particular may offer the opportunity for humans to guide their future genetic evolution. While most people see this as frightening and undesirable, it is possible that future generations will view genetic manipulation in human reproduction as a normal and helpful aspect of medical practice. It must be remembered, however, that none of us is genetically 'perfect': we all carry harmful recessive alleles. Some people agree that all the genetic variety of humankind should be valued and celebrated and that it would be wrong to aim for some kind of genetic correctness. We must use these new opportunities to change our genetic make-up wisely and cautiously, with the full understanding and consent of those involved and with careful consideration of the interests of unborn embryos and future generations.

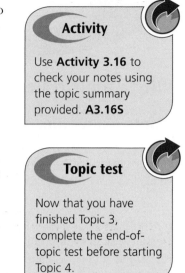

Activity

Use **Activity 3.16** to check your notes using the topic summary provided. **A3.16S**

Topic test

Now that you have finished Topic 3, complete the end-of-topic test before starting Topic 4.

Why a topic called Plants and climate change?

Plants – can't run, can't hide

It must be boring being a plant. You stay in the same place, grow, get old, lose your leaves and die. There's an exciting bit when you put out flowers and seeds but you don't do that very often and most seeds seem to be wasted.

Think what it would be like if like a plant, you had to stay in the same position. Walking on the moors, your foot gets caught between the rocks. Despite wriggling and crying for help, no one comes. It rains, the wind blows, the sun scorches, you are attacked by insects and other organisms. You become hungry and very thirsty. Within days you would be dead, whilst right next to you there are an enormous variety of plants surviving and multiplying. Who says humans are the dominant species? In such circumstances we are not fitted for survival and plants immediately command more respect.

Plants have to cope with any conditions their environment may throw at them. They can't run away or hide but they can perform processes that most animals can't. In this topic we shall look at how they do this, and what we can learn from them.

▼ **Figure 4.1** The giant redwood, *Sequoiadendron giganteum*, can grow to 90 m in height.

Plants can build the largest living structures (redwood trees, Figure 4.1). They produce the longest-lived single organisms (bristle cone pines, up to 4700 years old) and are the most prolific colonisers of the land (grasses). Indeed, about 97% of all the biomass on our planet is plant. Unlike animals they are solar powered. They use the products of photosynthesis to survive being fixed in one place.

People have utilised plant products throughout human history for building materials, clothes and also as fuel. We use seeds and parts of plants for food, adhesives and plastics. More recently we are learning to steal or copy their attraction and defence mechanisms so we can use them for pesticides, drugs and food additives. Plants don't do these things for our benefit. For them it is just a matter of 'survival of the fittest'. We take advantage of their ingenuity by selecting particular plants and cultivating them for our own use. This topic looks at how we exploit plants, and at how we are altering plants for our own use through breeding and genetic modification.

Climate change

Climate change will provide plants with a major problem. Individual plants rooted to one spot cannot move to escape the rising temperatures. Will some species become locally extinct as their distribution changes (Figure 4.2)? Will farmers switch to different crops to cope with changes in temperature and rainfall? Will new pests and diseases attack plants? In this topic we not only look at the possible consequences of climate change for plants but also the effects on the animals.

We consider whether climate change is really happening, what is causing any changes and how rising carbon dioxide levels might be reduced.

△ **Figure 4.2** Global climate change may lead in Britain to the disappearance of much-loved spring flowers, such as the bluebell (*Hyacinthoides non-scripta*).

Overview of the biological principles covered in this topic

This topic starts by looking at what is special about plants and at how humans rely on them. You will see how many of the central features of a plant's structure and lifestyle can be understood from a knowledge of plant cells.

Building on what you have studied in earlier topics, you will compare the structure and function of starch and cellulose, and contrast the structure of a typical plant cell with that of an animal cell. You will examine specialisation of cells for support and transport through the stem.

Unlike most animals, plants cannot move much – except when they produce pollen or seeds. You will look at how plant architecture and chemical defences allow them to cope with physical hardships and the risk of being eaten.

We use plants in countless ways – for foods, for clothing, for medicines and for much else besides. You will see that humans have relied on plants throughout our history and that new ways of using plants are still being devised that may make the use of resources more sustainable.

Throughout this topic you will see how scientists attempt to determine the possible impacts of global warming on natural ecosystems and agriculture. Building on the knowledge gained earlier in the course you will consider the effect of temperature on enzyme activity.

You will critically examine the different types of evidence used to support the theory that the Earth's climate is changing as a result of human activity. This leads on to a consideration of how scientific knowledge is produced and to the controversies that can arise if the predictions of science clash with politics and business. You will examine how scientists use models of climate change.

Recalling the basics of the carbon cycle studied at GCSE, you will gain a more detailed understanding of it and see how the use of biofuels and reafforestation can help reduce atmospheric carbon dioxide levels.

Review

Are you ready to tackle Topic 4 *Plants and climate change?*

Complete the GCSE review and GCSE review test before you start.

4.1 Big and strong

Why do trees grow tall? Some plants lift their leaves on stalks or trunks so that they are above those of their competitors. If it can grow big enough, a tree can obtain massive amounts of photosynthetic energy, and sink roots deep into the ground to take in the maximum amount of water and nutrients. If attacked by fungi, viruses or animals, part of their structure may be lost, but the rest can live on.

This is a risky strategy for survival since it can take a long time to grow to maturity. Yet it works; plants can build significantly bigger structures than any land animal. Redwoods can grow to 90 m, and 20 m palm trees can survive 100 mph winds (Figure 4.3). Such large structures must be strong enough to hold up their own weight and withstand the enormous wind forces on them. The same applies to upright annual plants; though they don't grow so tall because they only have one growing season in which to reach maturity.

◀ **Figure 4.3** Palm trees can withstand hurricane-force winds.

Wood is an amazing material

All plants use three basic principles to build tall structures:

1 They produce strong cell walls out of polymers made from sugar molecules.

2 Specialised cells build columns and tubes.

3 They stiffen these special cells with another polymer called lignin.

Trees add a ring of this stiffened (lignified) tissue each year, and building up wood in this way allows trees to grow taller. The structures produced are not only strong but also flexible, allowing trees to sway in high winds. We reproduce the properties of wood in composite materials, using more than one substance in combination. For example, concrete reinforced with steel is

Like trees, tall buildings have to be strong enough to withstand compression forces, so that they do not collapse under their own weight. Materials such as concrete provide strong walls that will bear this weight.

Tall buildings also have to deal with the horizontal force of wind. At their tops, most skyscrapers can move several feet in either direction, like a swaying tree, without damaging their structure. Any more movement would not only be uncomfortable for people inside, but would risk breaking up the concrete structures – these are brittle, unable to bend much without breaking.

Steel columns allow a degree of bending – they are much less brittle than concrete. However, a tall building made entirely from steel columns would eventually buckle and bend under its own weight. The solution is a steel framework combined with a concrete structure, providing both strength and stiffness (Figure 4.4).

A tall building needs strong walls and a frame of columns and beams to support the structure. The building must not be allowed to sway too much, so it is necessary to bolt and weld the horizontal beams to the vertical structure. This makes the entire structure move as one unit, like a tree.

Figure 4.4 Buildings have to withstand the same forces as tall trees. Architects copy plants when designing buildings using concrete walls reinforced with steel instead of cellulose walls reinforced with lignin.

much stronger than either material alone. This is why many modern buildings and other structures are still made of wood, itself a composite material.

In order to appreciate how plant cells are highly specialised for their function you first need to study a typical plant cell.

Key biological principle: How are plant cells different from animal cells?

Figure 4.5 on page 148 shows the ultrastructure of a generalised plant cell. Decide how it is different from the animal cell shown in Figure 3.8 (page 100).

There are two fundamental differences:
• The plant cell has a rigid cell wall.
• It contains chloroplasts.

Chloroplasts are the site of photosynthesis, where solar energy is converted to chemical energy. Starch is found in storage vacuoles called amyloplasts in the cytoplasm. In addition to the cell wall and chloroplasts in plant cells, there is often a large central vacuole surrounded by a vacuolar membrane (tonoplast).

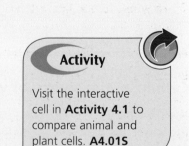

Activity

Visit the interactive cell in **Activity 4.1** to compare animal and plant cells. **A4.01S**

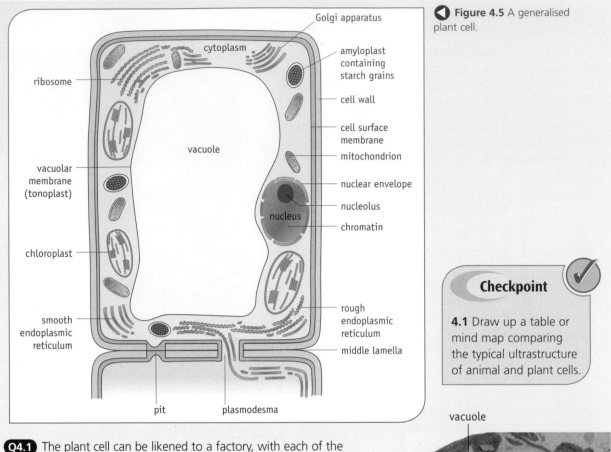

Checkpoint ✓

4.1 Draw up a table or mind map comparing the typical ultrastructure of animal and plant cells.

Q4.1 The plant cell can be likened to a factory, with each of the organelles carrying out a special function. Look at factory functions below and decide which of the cell organelles perform an equivalent function:
a control centre where the instructions are kept to tell the factory floor what to make
b boiler room, harnessing energy to drive all the processes in the factory
c packaging hall, wrapping up and collecting products to be exported from the factory
d security screen, allowing in only those external raw materials which the factory needs
e warehouse storage facility within the factory.

Parenchyma is a type of plant tissue found throughout the plant. The cells fill spaces between more specialised tissues and may themselves have certain specialised functions. For example, in roots they may have a role in storage, and in leaves they contain chloroplasts and form the photosynthetic tissue (Figure 4.6).

▲ **Figure 4.6** False colour transmission electron micrograph showing parenchyma cells in the leaf containing many chloroplasts.

vacuole
starch granules
nucleus
chloroplasts

Cellulose, the source of wood's strength

Wood's strength comes in part from the thin **cellulose** walls of plant cells and the 'glue' that holds them together.

Cellulose is a polysaccharide. It is a polymer of glucose, but the glucose it is made of is slightly different from that which forms starch.

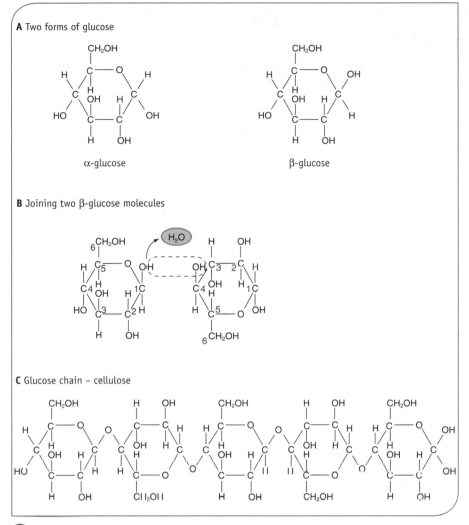

Figure 4.7 A Cellulose is made up of β-glucoses joined together. Starch is composed from α-glucoses. **B** Formation of a 1,4 glycosidic bond between the two β-glucose molecules is only possible if one molecule is rotated through 180°. **C** A molecule of cellulose. Notice how the –OH groups project from both sides of the molecule due to the arrangement of the glucose molecules. Each alternate glucose is inverted to allow the 1,4 glycosidic bond to form.

In Figure 4.7 the two forms of glucose are shown alongside each other. Can you spot the difference? You should notice that the –OH (hydroxyl) groups on the first carbon atoms are on opposite sides.

Cellulose is made up of β-glucose units. A condensation reaction between the –OH group on the first carbon of one glucose and the –OH on the fourth carbon of the adjacent glucose links the two glucose molecules. A 1,4 glycosidic bond forms. In cellulose, all the glycosidic bonds are 1,4; unlike starch, there are no 1,6 glycosidic bonds. Because of this, cellulose is a long unbranched molecule.

Q4.2 Look at Figure 4.7B. Describe the relative positions the two β-glucose molecules must have for a condensation reaction to take place between them.

Q4.3 Explain why the bond between each pair of glucose molecules in cellulose is called a 1,4 (glycosidic) bond.

Support

You can remind yourself about condensation reactions and –OH groups by looking at the Biochemistry support on the website.

Looking at a pair of nylon tights under the microscope shows that there are tight chains running around the leg of the tights. These chains are joined more loosely vertically, so there is more 'give' lengthways. Net bags and elastic tubular bandages also show this uneven stretch. It is cross-links between the units that determine the amount of movement, and so the strength in any direction. Cross-linkages also strengthen scaffolding, electricity pylons and steel girders (Figure 4.8).

On a molecular level, cross-linking gives strength and stiffness to materials such as plastics. There are two basic types of plastic.

- A plastic in which the polymer molecules are not cross-linked is a thermoplastic. Since the molecules are not connected, they can move apart when the plastic is heated, so the plastic will soften, melt or flow when you heat it.
- In a thermoset plastic, the polymer molecules are cross-linked by heating during production. A thermoset plastic does not soften, melt or flow when heated. Thermoset plastics can be used for products such as kettles which will be heated. They are much harder and more rigid than thermoplastics, and can replace metal in structures such as window frames, sinks and water pipes.

△ **Figure 4.8** This structure relies on cross-linking to provide the structure with strength.

Microfibrils, bundles of cellulose molecules

Each cellulose chain typically contains between 1000 and 10 000 glucose units. Unlike an amylose molecule that winds into a spiral (Figure 1.38, page 35), the cellulose molecules remain as straight chains. Hydrogen bonds form between the –OH groups in neighbouring cellulose chains, forming bundles called **microfibrils** (Figure 4.9). Individually, the hydrogen bonds are relatively weak compared with the glycosidic bonds, but together the large number of hydrogen bonds in the microfibril produce a strong structure.

Cell walls – the secret of their strength

If we look more closely at the cell wall (Figure 4.10), we can see that it is formed of microfibrils. These microfibrils are bundles of about 60–70 cellulose molecules. The microfibrils are wound in a helical arrangement around the cell and stuck together with a polysaccharide glue. Figure 4.10 shows how successive layers of the microfibrils are laid down at angles to one another, forming a composite structure.

The glue that holds together the microfibrils is composed of short, branched polysaccharides known as **hemicelluloses** and **pectins**. These short polysaccharides bind both to the surface of the cellulose and to each other, and hold the cellulose microfibrils together.

Pectins are also an important component of the middle lamella – the region found between the cell walls of adjacent cells. The pectins act as cement and hold the cells together.

Activity

In **Activity 4.2** you can use the Biochemistry support on the website to see how β glucoses join to form cellulose molecules, and also to compare the structures of starch and cellulose. **A4.02S**

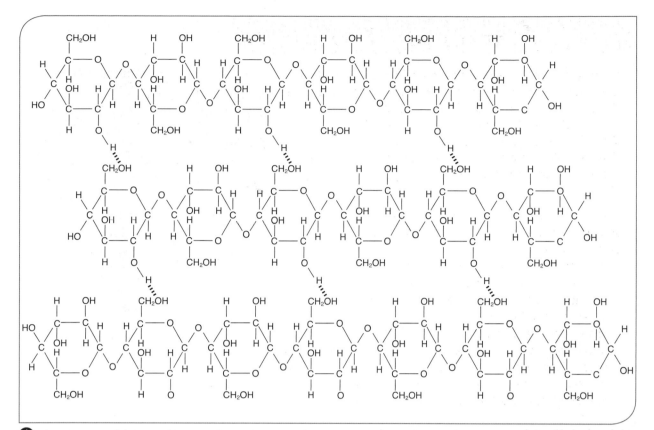

🔺 **Figure 4.9** In cellulose, neighbouring chains of glucose molecules are linked by hydrogen bonds to form microfibrils.

The arrangement of the cellulose microfibrils within a matrix of hemicelluloses makes the cell wall very strong – rather like steel-reinforced rubber tyres. In this analogy, the hemicellulose matrix is the rubber, and the cellulose microfibrils are the reinforcing steel cables. This makes a strong but pliable structure. The microfibrils are laid down at different angles, which makes the wall strong and flexible.

🔺 **Figure 4.10** In the cell wall, layers of cellulose microfibrils are laid down at different angles.

Did you know? Composite materials

The use of composite materials is nothing new. Ancient Egyptians put cut straw into their bricks, and Incas and Mayas put fibres into their pottery. The addition of fibres to a brittle matrix stops cracks, so the material is less likely to fracture. The result is a much stronger material. Papier mâché was used for mummy cases, and more recently for aircraft parts during the Second World War.

A boat or furniture built out of glass alone would be brittle, but when glass fibres are set in a bonding resin, the resulting glass-reinforced plastic (fibreglass) is strong, tough and water resistant.

Reinforcement of concrete with iron or steel bars or meshes allows concrete to be used in situations where it bends, for example in a beam. This greatly increases the versatility of concrete.

Checkpoint

4.2 Look back at the structure of starch in Topic 1. Draw up a table comparing the structures and functions of starch and cellulose.

Did you know? Pectins, jams and fruit juice

When fruits ripen they usually become soft (and eventually squashy) as the pectins holding the cells together are broken down by enzymes. The pectins become soluble, allowing the cells to shift when you squeeze the fruit.

In the fruit juice industry pectins are a nuisance. They thicken the juice, making it difficult to extract from the fruit, and slowing down filtration. Enzymes called pectinases are used commercially to break down the pectins, increasing yield and giving juices

such as apple juice their clarity. Pectinases are also used to make baby food purées, and to peel the citrus fruits that come in tins.

Conversely, when making jam, pectin is a good thing – the more pectin, the more solidly the jam sets, and the less fruit you have to use per jar. But if the fruit is too ripe the natural pectins will have broken down, and the jam will remain runny unless more pectin is added.

Crossing the cell wall

The cell wall does not separate plant cells completely. Narrow fluid-filled channels, called **plasmodesmata**, cross the cell wall, making the cytoplasm of one cell continuous with the cytoplasm of the next. The cell wall is also fully permeable to water and solutes. Can you identify the plasmodesmata in the cell wall shown in Figure 4.11? (Look at the photograph and decide before reading the caption.)

At some places the cell wall is thin because only the first layer of cellulose is deposited. The result is a **pit** in the cell wall. Plasmodesmata are often located in these pits, aiding the movement of substances between cells.

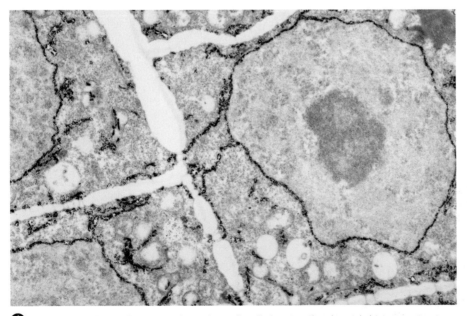

Figure 4.11 Photo of a section through a cell wall showing the channels between adjacent cells, the plasmodesmata. These contain tubes that link the endoplasmic reticulum of neighbouring cells.

Tubes for transport and strength

To build a tall plant such as a tree, some of the cells within the stem must be stiffened to provide mechanical support. At the same time, some cells must allow water and minerals (inorganic ions) to pass from the roots to the leaves.

There are two specialised types of cell of particular importance in fulfilling these functions. These are:

- xylem vessels – these form tubes for transport, and their stiffened cell walls help support the plant
- sclerenchyma fibres – columns of these cells with their stiffened cell walls also provide support.

Where in the stem are these specialised cells?

There are three basic types of tissue found within plants. Figure 4.12 shows the general location of these three types: dermal tissue (epidermis), vascular tissue and ground tissue.

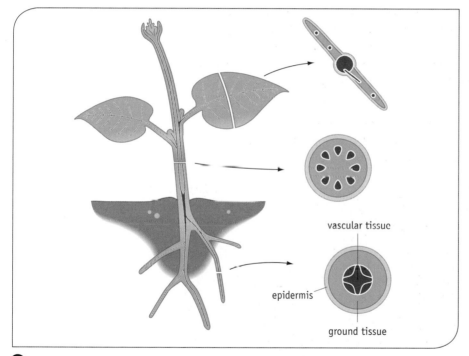

vascular tissue

epidermis

ground tissue

Figure 4.12 Where are the three basic types of tissue in the plant? The epidermis is a single layer of cells covering the entire outside of the plant. The vascular tissue, involved in transport, is surrounded by the ground tissue which contains cells specialised for photosynthesis, storage and support.

Figure 4.13 shows the location of the **vascular tissue** within a cross-section of a stem from a dicotyledon (Figure 4.14). Notice how each **vascular bundle** contains **xylem vessels** and **phloem** sieve tubes. On the outside of the bundle are located **sclerenchyma fibres**. In a young dicotyledon, the vascular tissue is in bundles towards the outside of the stem. As the plant ages, these separate bundles merge to form a continuous ring. The xylem vessels carry water and inorganic ions up through the stem; the phloem transports sugars made by photosynthesis in the leaves up and down the plant.

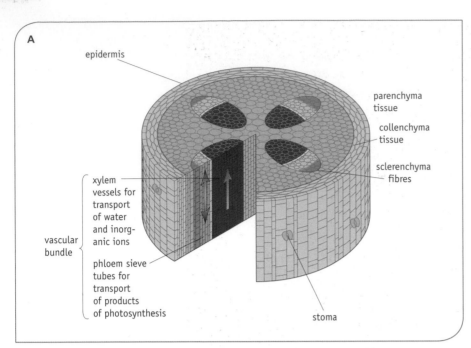

A

epidermis

parenchyma
tissue

collenchyma
tissue

sclerenchyma
fibres

xylem
vessels for
transport
of water
and inorg-
anic ions

vascular
bundle

phloem sieve
tubes for
transport
of products
of photosynthesis

stoma

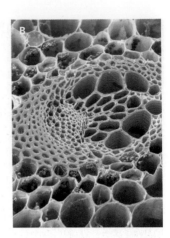

B

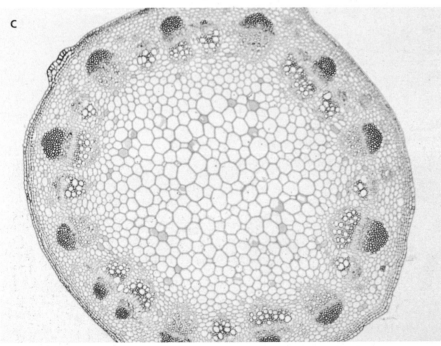

C

⚠ **Figure 4.13 A** The general arrangement of vascular tissue within the stem of a dicotyledon. The arrows represent the direction of flow through the xylem and phloem.
B Coloured scanning electron micrograph of a vascular bundle of a buttercup, *Ranunculus repens*. The phloem sieve tubes are orange/yellow with the sclerenchyma fibres on their left. Magnification ×125. **C** A photomicrograph of a transverse section of a sunflower, *Helianthus*, stem.

Q4.4 Identify the location of: **a** a vascular bundle **b** xylem vessels **c** sclerenchyma fibres in Figure 4.13C.

Activity

In **Activity 4.3** you can examine plant stems and locate the different tissue types yourself.
A4.03S

▲ **Figure 4.14** Flowering plants, the angiosperms, can be divided into two classes depending on the number of seed leaves they have. Monocotyledon embryos have one seed leaf, their leaves have parallel veins and they rarely grow large because they cannot produce true wood. Dicotyledon embryos have two seed leaves and the veins on their leaves form a network pattern. They can produce wood and so can form trees and shrubs. The grass on the left is a monocotyledon. The clover is a dicotyledon.

Did you know? Strong tubes

Tubes are useful things. They can carry materials, and because they are hollow they are light. But tubes have a tendency to collapse in on themselves when put under pressure. So they need to be reinforced if they are not already made of a strong material (Figure 4.15).

The supporting cells in plants are tubular, and are strengthened with spirals and rings of ligin.

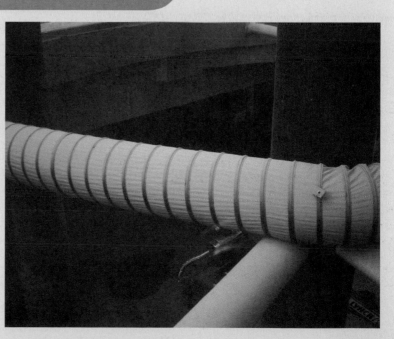

▶ **Figure 4.15** This large-scale vacuum tube uses spiral thickening to hold it open.

Xylem vessels for transport

Figure 4.13A shows a cross-section of a plant stem. The xylem vessels are made up of large cells with thick cell walls. They form a column of cells acting as pipes for the transport of water and inorganic ions. In order to transport water, the cell walls have to be waterproofed. The plant does this by producing another polymer – **lignin**. This polymer impregnates the cellulose cell wall and as the cells become lignified, the entry of water and solutes into them is restricted. At about the same time the tonoplast breaks down and there is autolysis of the cell contents. The cell organelles, cytoplasm and cell surface membrane are broken down by the action of enzymes and are lost. The cell dies.

The detailed structure of xylem vessels is shown in Figure 4.16. The end walls between the cells of the column are lost or become highly perforated. Long tubes form as a result of this process which are continuous from the roots of the plant to the leaves. These tubes transport water and minerals upwards through the plant. The cellulose microfibrils and the lignin in the cell walls of the xylem vessels give the tubes great strength.

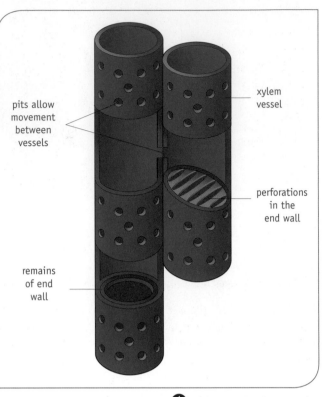

▲ **Figure 4.16** Xylem vessels cut open to see their detailed structure.

Q4.5 Explain how disruption of the tonoplast might result in the breakdown of the cell contents.

How is water transported through xylem vessels?

Xylem vessels are effectively fluid-filled tubes through which water moves upwards from the roots to the shoots. How is the water moved through these dead cells?

Water evaporates from all surfaces of the plant, mostly from the large surface area of the leaves. The majority of the evaporation occurs from the surfaces of the cells that line the substomatal cavities in the leaves, as shown in Figure 4.17. Water diffuses out through the stomata down a **diffusion gradient**. Water evaporating from the plant in this way is known as **transpiration**. The water that leaves a plant leaf by transpiration is replaced by water absorbed through the roots. In the tallest trees water may move up to 100 m through the plant.

The loss of water from the cells that border the substomatal cavity causes the concentration of water in them to fall. As a result, water enters from the adjacent cells. Water either diffuses through the plasmodesmata between cells or diffuses along the cellulose cell walls. It can also move by osmosis, passing across the cell surface membranes and tonoplasts.

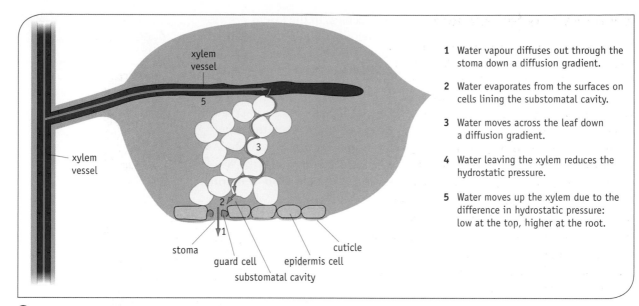

1 Water vapour diffuses out through the stoma down a diffusion gradient.

2 Water evaporates from the surfaces on cells lining the substomatal cavity.

3 Water moves across the leaf down a diffusion gradient.

4 Water leaving the xylem reduces the hydrostatic pressure.

5 Water moves up the xylem due to the difference in hydrostatic pressure: low at the top, higher at the root.

▲ **Figure 4.17** Water moves up through the xylem due to the pressure gradient.

Water moving out of the cells of the leaf is replaced by water from nearby xylem vessels. Removing water from the xylem vessel reduces the hydrostatic pressure (pressure exerted by the liquid in all directions) within the vessel. This makes the hydrostatic pressure at the top of the vessel lower than that at the bottom. This difference in pressure causes water to be drawn up the tube, moving from high hydrostatic pressure to low hydrostatic pressure. Water moves up through the xylem vessel as a continuous column; this movement of water is known as the transpiration stream.

A continuous column of water

Two forces – cohesion and adhesion – help to hold the water in the xylem as a continuous column (Figure 4.18). These forces are the result of the polarity of the water molecules, see page 158.

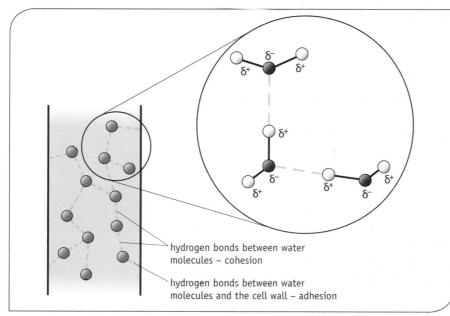

hydrogen bonds between water molecules – cohesion

hydrogen bonds between water molecules and the cell wall – adhesion

Activity

The interactive tutorial in **Activity 4.4** should help you understand how water moves through the stem. Use the Biochemistry support on the website to remind yourself about water and hydrogen bonding. **A4.04S**

◀ **Figure 4.18** A sketch to show the forces of cohesion and adhesion in a xylem vessel.

Adhesion is the attraction between unlike molecules. In the xylem vessels the water adheres to the walls of the vessels, helping to hold the column of water within the vessel. The narrow diameter of the vessels means that there is a large surface area to volume ratio within the tubes, thus creating high adhesive forces.

Cohesion is the attraction between like molecules. There are strong cohesive forces between water molecules. In the xylem vessel these keep the water together as a continuous column, preventing it breaking along its route. As water leaves the top of the xylem, the cohesive forces ensure that the whole column of water moves upwards. This is known as the **cohesion–tension theory**. The thickening of the xylem walls prevents the vessels collapsing under the tension that builds up in the xylem.

The rate of transpiration for a plant will vary, transpiration generally being faster when conditions are warm and dry, increasing evaporation from the leaf surface. On the other hand, transpiration rates are low when the plant is in the dark or is short of water, when the stomata on the leaf surfaces close to conserve water.

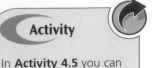

Activity

In **Activity 4.5** you can investigate transpiration in trees. **A4.05S**

What else do xylem vessels transport?

The movement of water through the xylem provides a **mass flow system** for the transport of inorganic ions. These are absorbed into the roots and required throughout the plant. For example, nitrates are absorbed by the plant and used in every cell to manufacture proteins, nucleic acids, ATP and plant growth substances.

If inorganic ions are not absorbed in sufficient amounts, the plant will show deficiency symptoms. For example, if the plant lacks magnesium it is unable to make chlorophyll and the older leaves become yellow (yellow leaves may also be a sign of nitrogen deficiency). A lack of calcium causes stunted growth due to the role of calcium ions in the structure of the cell wall and in the permeability of the cell membrane.

Q4.6 There is a very low concentration of inorganic ions in the soil solution surrounding roots. Using what you know about transport across cell surface membranes, suggest how the inorganic ions are transported into the root cells.

Xylem and sclerenchyma for support

Lignin not only waterproofs the cell wall, it also makes it much stiffer and gives the plant much greater tensile strength. Instead of forming a uniform layer on the inside of the xylem cell walls, lignin is often laid down in spirals or in rings as shown in Figure 4.19.

Xylem vessels are not the only cells that become impregnated with lignin. The sclerenchyma fibres that are associated with vascular bundles in the stem and leaves also have lignin deposited in their cell walls. Sclerenchyma comes from the Greek 'scleros' meaning 'hard'. As with xylem vessels, the sclerenchyma fibres die once lignified, leaving hollow fibres (Figure 4.20). The strength of these fibres varies in different plant species depending on the length of the fibres and the degree of lignification.

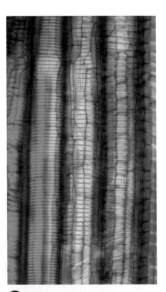

▲ **Figure 4.19** The lignin thickening in xylem vessels can take different forms as shown in this photomicrograph.

Key biological principle: The importance of water

Water, H_2O, is unusual among small molecules. It is a liquid at 'normal' biological temperatures; at room temperature most other substances made of small molecules, such as CO_2 and O_2, are gases. Water is a very small polar molecule and it is this polarity that accounts for many of its biologically important properties.

Why is water polar?

Each of the two hydrogen atoms in a water molecule shares an electron pair with the oxygen, so forming a covalent bond. The unshared pairs of electrons repel the shared pairs so the two hydrogens are pushed slightly towards each other, forming a V-shaped molecule (Figure 4.18 on page 157). As a result, the hydrogen end of the molecule is slightly positive and the oxygen end is slightly negative because the electrons are more concentrated at the oxygen end.

The positively charged end of a water molecule is attracted to the negative ends of surrounding molecules. This hydrogen bonding holds the water molecules together and results in many of the properties of water, including its being liquid at room temperature.

Solvent properties

Many chemicals dissolve easily in water allowing vital biochemical reactions to occur in the cytoplasm of cells. Free to move around in an aqueous environment the dissolved chemicals can react, often with water itself being involved in the reactions, for example in hydrolysis and condensation reactions. Dissolved substances can also be transported around organisms, in plants through the xylem and phloem, and in animals via the blood and lymph systems.

Ionic substances, such as sodium chloride (NaCl), dissolve easily in water. In the case of sodium chloride, the negative Cl^- ions are attracted to the positive ends of the water molecules while the positive Na^+ ions are attracted to the negative ends. The chloride and sodium ions become hydrated in aqueous solution, that is, surrounded by water molecules.

Polar molecules also dissolve easily in water. Their polar groups, for example the –OH group in sugars or the amine group, –NH, in amino acids, become surrounded by water and go into solution. Such polar substances are said to be hydrophilic – 'water-loving'.

Non-polar, hydrophobic, substances, such as lipids, do not dissolve in water. As we saw in Topic 2, this property of lipids is important in maintaining the structure of cell membranes.

Thermal properties

he amount of energy in joules required to raise the temperature of 1 g of water by 1 °C, is very high. This is because in water a large amount of energy is required to break hydrogen bonds. A large input of energy causes only a small increase in temperature so water warms up and cools down slowly. This is extremely useful for organisms, helping them to avoid rapid changes in their internal temperature and enabling many of them to maintain a fairly steady temperature even when the temperature in their surroundings varies considerably.

Density and freezing properties

Unlike most liquids, water expands as it freezes. As liquid water cools the molecules slow down, enabling the maximum number of hydrogen bonds to form between the water molecules. These hydrogen bonds hold the water molecules further apart than in liquid water, making ice less dense than liquid water. So ice floats, enabling organisms to survive in liquid water under the ice in frozen ponds and lakes.

Cohesion and surface tension

Hydrogen bonds hold water molecules together and as we have seen, these cohesive forces are extremely important for organisms, for example in the transport of water up through xylem vessels.

Surface tension at water surfaces is also due to these cohesive forces between water molecules, causing the surface layer of water to contract. This is useful for some small aquatic organisms, such as duckweed and pond skaters, which can be supported on this surface film.

The taller a plant needs to grow, the greater the proportion of its stem that becomes lignified. In a tree, this is the majority of the trunk: the living parts are towards the surface (under the bark) and grow new layers each year. These layers are the annual rings. In annual plants there is relatively little lignification. The plant stem must rely on tightly packed, fully turgid parenchyma cells to maintain its shape and keep it erect. A **turgid** cell is one that is completely full, with its cell contents pressing out on the cell wall. If a cell loses water, turgor is lost. If a high proportion of a plant's cells lose their turgidity, the plant will wilt.

Useful plant fibres

The potential for plants to benefit humans is enormous. People have been using plant fibres for thousands of years in order to make products such as clothing, rope, floor coverings, paper and many more. Plant fibres can be used in these ways because they are:

- long and thin
- flexible
- strong.

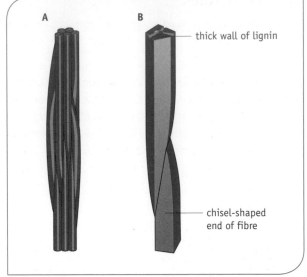

▲ **Figure 4.20 A** Several sclerenchyma fibres. **B** Part of two fibres shown at a higher magnification.

How do we extract fibres from plants?

To obtain fibres we must take the plant apart. Fortunately, cellulose – and particularly cellulose combined with lignin – is very resistant to chemical and enzymic degradation, whilst the polysaccharides that hold the fibres together can be dissolved away.

The more lignin is present, the harder it is to separate fibres. So to produce fibre pulp from trees, caustic alkali is required. For flax (Figure 4.21) and other suitable plants a milder treatment is used and in some traditional processes the stems are piled in heaps, allowing bacteria and fungi to do the work. This process, and its more modern chemical and enzymic equivalents, is called 'retting'.

Q4.7 What do you think makes plant fibres a success commercially?

▲ **Figure 4.21** Scanning electron micrograph of fibres from a flax plant (*Linum usitatissimum*) used in linen fabric. Magnification ×3150.

Not just textiles ...

Fibres have many other uses in addition to the textile industry. For example, mats of fibre are used to absorb heavy metals and also hydrocarbons from polluted water (Figure 4.22).

Plant fibres can also be added to other materials to form biocomposites. For example, when oil seed rape fibres are mixed with plastic the resulting material is stronger than plastic alone. These biocomposites are also more renewable, more biodegradable and can be easier and safer to handle than artificial fibres. Researchers in Australia recently announced that they thought a car made entirely from plant material was possible within 10 years. They were a little behind the times: Henry Ford produced a car made entirely from hemp in 1941.

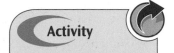

Checkpoint

4.3 Look back over all the information on xylem vessels and sclerenchyma fibres. Then list the physical properties that enable them to be used for human benefit.

Activity

In **Activity 4.6** you can have a go at extracting nettle fibres, using retting, and then testing their strength. **A4.06S**

▲ **Figure 4.22** Oil pollution control. Natural fibres absorb hydrocarbon pollution from water.

4.2 Repel your enemies

Chemical defences against attack

A plant can't move much, it can't run and it can't hide, so it is an easy target for animals, bacteria and fungi. To try and avoid being the next passing herbivore's lunch, plants have chemical defences which repel and even kill their predators. One strategy is to produce a chemical that is distasteful or even toxic. If the animal takes a bite and the taste is offensive (usually bitter, Figure 4.23) the animal is deterred from feeding further. If its chemicals kill the predator the plant will survive. A classical example is provided by a plant called pyrethrum.

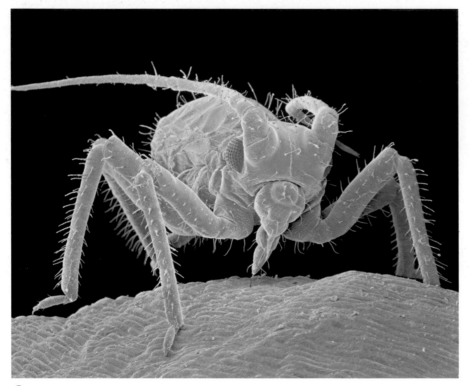

▲ **Figure 4.23** The bitter taste in tea is caused by the presence of tannins in the leaves. These also have the effect of disruption of digestion in aphids and so protect the tea plant from this form of insect attack. Magnification ×105.

Natural antibacterials

Plants sometimes store toxic compounds in hairs on the surfaces of their leaves. This is very obvious in the stinging nettle, which most of us avoid, but not so apparent in mint which produces the chemicals menthol and carvone. These chemicals are toxic to microbes and some insects but attractive to us as flavouring in foods or tea. Even humans experience the numbing effect of mint (whilst brushing your teeth, for example) but to microorganisms this can be lethal.

Garlic extracts have been found to destroy bacteria such as *Campylobacter* and *Helicobacter* which cause intestinal infections. This is potentially

Did you know? Pyrethrum – a natural insecticide

Pyrethrum is a kind of chrysanthemum and a member of the daisy family (Figure 4.24). Pyrethrum has the remarkable property of having no known pests or diseases. It contains chemicals which are extremely toxic to insects. Pyrethrum is grown commercially in Kenya, where flowers are picked, dried in the sun, then bagged and transported to factories. Here they are crushed into powder and further refined to extract the active ingredient: various pyrethrins.

Pyrethrins are used in many insecticide recipes for sprays on fruit and flowers, flea treatments for cats and dogs, and the reduction of tick infestation in cattle in their native home, Kenya.

Fortunately, these compounds are much less toxic to vertebrates than to insects, but they are not entirely harmless and must be handled with care. Pyrethrins have the advantage that they are unstable after spraying and rapidly decompose to harmless residues.

You might try growing some chrysanthemums and testing them on flies.

One alternative to using pyrethrins is to use the plants themselves. Organic gardeners use chrysanthemums and marigolds as 'companion' plants – planted amongst vegetables they act as a natural insect repellent.

▲ **Figure 4.24** *Tanacetum coccinium*, the plant from which we get pyrethrum.

important as some strains of the bacteria are now resistant to widely used antibiotics such as penicillin. The active ingredient in garlic is allicin; this is known to interfere with lipid synthesis and RNA production. Allicin is only produced when the plant is cut or damaged. Its inactive precursor, alliin, is converted into the active form by the enzyme alliinase.

Some studies have shown that certain parts of a plant tend to have greater antibacterial properties than the rest. These are typically the seed coat, fruit coat, bulb and roots.

Q4.8 Suggest why the parts of the plant listed above need to produce the most antibacterial chemicals.

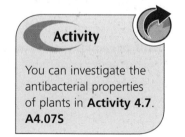

Activity

You can investigate the antibacterial properties of plants in **Activity 4.7**. **A4.07S**

Medicines from plants

Not surprisingly, many plants contain these poisons, or produce them rapidly as a response to wounding. However, 'poison' is a relative term and relates to the dose necessary to kill an organism. Clearly, if a chemical can kill pathogenic (disease-causing) microbes or malignant cancer cells at a dose level which leaves humans alive, then this 'poison' is a likely medicine. Not surprisingly, an enormous number of medicines are derived from plants.

The World Health Organization estimates that 75–80% of the world's population uses plant medicines either in part or exclusively. Many of the common medicines we use were originally plant derived, such as aspirin (derived from an extract of willow bark), salicylic acid, and morphine and codeine (both derived from opium poppies).

Digitalis and drug development

Foxgloves and dropsy

Foxglove leaves are poisonous when eaten by humans and other animals. They have a strong, bitter taste which serves as a warning. The symptoms of poisoning are dizziness, vomiting, hallucination, and heart failure caused by an irregular heartbeat. But it is the effect on the heartbeat that made it a traditional folk remedy when used in moderation.

The foxglove (Figure 4.25) was known for centuries to have medicinal qualities and in particular was used to treat a condition known as dropsy. Dropsy, now called oedema, happens when fluid accumulates in the body tissues (see Topic 1, page 28). This process is painful and can cause a slow death. Oedema is usually caused by heart or kidney problems. A fast and irregular heartbeat is one of the signs.

▲ **Figure 4.25** The foxglove, *Digitalis purpurea*, has been used in medicines for hundreds of years.

Because the blood pressure is raised, tissue fluid fails to return to the capillaries and accumulates in the patient's feet, legs and other organs, causing them to swell up. Eventually the patient may drown as fluid fills up their lungs. So it was not surprising that a herb known to relieve this serious condition was used as a remedy. However, it was not until a country doctor called William Withering (Figure 4.26) published *A Treatise on the Foxglove* in 1775 that it became an accepted form of medicine.

Withering had heard of the foxglove's curative properties for dropsy, but his attention was focused when he met Mrs Hutton, a 'wise woman' who was showing signs of the disease. She assured him that she would be all right after she'd had a cup of her special tea. Imagine Withering's surprise when he visited her again and found that she had recovered.

Talking to Mrs Hutton, Withering found out that foxglove was amongst the 20 odd herbs used in the potion. He suspected that something in the foxglove was the active ingredient. Later Mrs Hutton sold her recipe to Withering and he began to investigate the plant further. One of his first patients was a brewer who was suffering from swollen limbs and an irregular heartbeat. After a few doses of Withering's 'digitalis soup' he became healthy and his pulse became 'more full and regular'. Unfortunately his next patient, an old woman, nearly died with the treatment so Withering gave up his investigations.

WILLIAM WITHERING, OLD F.R.S.
FELLOW OF THE LINNEAN SOCIETY

▲ **Figure 4.26** William Withering.

Getting the dose right

After moving to Birmingham General Hospital, which received many patients suffering from dropsy, Withering was persuaded to renew his investigations. Studying 163 patients, he discovered and recorded side effects of digitalis. These included nausea, vomiting, diarrhoea, and a rather strange green/yellow vision. A sign that the patient was recovering was the production of a large quantity of urine.

Withering realised that getting the dose right for the patient was of vital importance. He applied a standard procedure to discover the correct dosage for each patient. He would slowly increase the dose until the patient started to have diarrhoea and vomiting, and then reduce it slightly; this would be the most effective dose. Withering meticulously recorded all his results, and after 10 years wrote his book about the medicinal properties of the foxglove. This helped change the face of medical practice forever. We now know that the active ingredient in the foxglove is a single chemical called digitalin.

 Q4.9 List the sequence of steps that Withering took when testing his drug.

Drug testing today

Today a potential new drug must pass a series of tests if it is to be developed into a new product. It has to be proven to be effective, safe and capable of making a profit. It can typically take 10–12 years and cost about US$500 million to develop a drug.

Potential substances are analysed and the active ingredient (the compound which may bring about a cure) is identified and copied so that it can be manufactured synthetically. Slight variations of the chemical structure are made just in case they might have a better effect.

A series of trials of the compound now begins. There are four stages in all. The first is pre-clinical testing, followed by three phases of clinical trials. Table 4.1 describes these four stages.

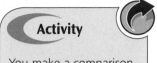

Activity

You make a comparison between William Withering's approach to drug development and that of the drug companies in **Activity 4.8**. **A4.08S**

▼ **Table 4.1** Stages in drug testing.

Pre-clinical testing	Laboratory and animal studies assess safety and determine whether the compound is effective against the target disease. These tests can take several years to complete. Thousands of chemicals go through pre-clinical testing but only a handful are ever approved for clinical trials on humans.
Clinical trials – phase I	A small group of healthy volunteers are told about the drug and given different doses. The trial confirms whether or not the compound is being absorbed, distributed, metabolised and excreted by the body in the way predicted by the laboratory tests. The effects of different doses are monitored. In the UK a review of the data collected is made by the UK Medicines Control Agency, an independent body of scientists and doctors. The agency recommends whether or not to proceed to phase II trials.
Clinical trials – phase II	Small groups of volunteer patients (100–300 people with the disease) are treated to look at the drug's effectiveness. If the results are promising, phase III trials are set up.
Clinical trials – phase III	A large group of patients (1000–3000 people) is selected and divided randomly into two groups. One group is given the compound that is being investigated and the second is given an inactive 'dummy' compound known as a **placebo**. It is important that neither the patients nor the doctors know who is having the compound under investigation and who is having the placebo; this is known as a **double-blind test**. If the compound being investigated is effective, then the results will show a statistically significant improvement in the patients receiving the treatment compared with patients given the placebo. The tests also look for any adverse reactions by the patients. The way is now open to license the compound as a drug, after which it can be marketed.

Q4.10 **a** Explain why the review team in the phase I trials needs to be independent.
b If phase II is a success, why is phase III needed?
c Is it ethical to give some patients a placebo, when it is known that the compound is likely to have a beneficial effect?
d Sometimes patients on a placebo will show an improvement in their condition. Suggest why this may be so.

4.3 Try again next year

Seeds for survival

Having successfully survived the onslaught of all sorts of herbivores, the plant's next challenge is to make sure enough of the next generation also survive. Flowering plants have achieved this by packaging a miniature plant in a protective coat with its own food supply; we call them seeds. Inside the seed the embryo remains dormant until conditions are suitable for restarting growth (Figure 4.27).

Most seeds can remain viable (still able to germinate) in the soil for extended periods of time. For some seeds it may only be a year or two but for other species it can be much longer. Seeds taken from dated burial sites have been successfully germinated after hundreds of years. As a rough generalisation, 80% of cereal seeds (e.g. wheat) will remain viable for 15 years if dehydrated and stored at –20 °C.

In the wild, the presence of seeds of any one species within the soil is known as that species' seed bank. A seed bank will have seeds from several different years. If the adult plants of a particular species in an area fail to survive or reproduce, then the seeds remaining in the seed bank may germinate and allow the species to survive there.

Seeds are vital to the survival of a plant. They:
- protect the embryo
- aid dispersal
- provide nutrition for the new plant.

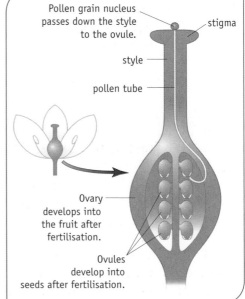

Figure 4.27 The seed structure ensures that the embryo it contains has the best chance of germinating and establishing a new plant.

What's in a seed?

In flowering plants the ovule is fertilised by the nucleus from a pollen grain and develops into the seed. You can locate the position of the developing seeds in the generalised flower shown in Figure 4.28.

The embryo develops into three distinct parts – a **radicle** (young root), a **plumule** (young shoot) and one or two **cotyledons** (seed leaves). These different parts can be seen in Figures 4.29 and 4.30.

In some species the stored food in the seed remains outside the embryo in storage tissue called **endosperm**, as shown in Figure 4.29. This is common in monocotyledons, for example cereals. Seeds of this type are called endospermic. In many dicotyledons the embryo absorbs the stored nutrients from the endosperm and the food is stored in the seed leaves (cotyledons) which swell to fill the seed (Figures 4.30 and 4.31).

Pollen grain nucleus passes down the style to the ovule.

stigma

style

pollen tube

Ovary develops into the fruit after fertilisation.

Ovules develop into seeds after fertilisation.

 Figure 4.28 Seeds develop inside the ovary.

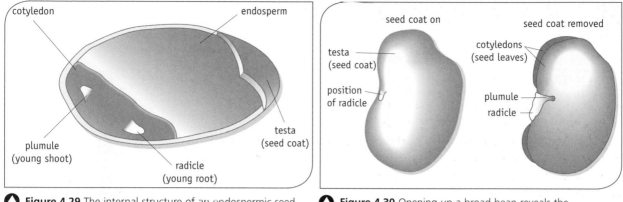

Figure 4.29 The internal structure of an endospermic seed.

Figure 4.30 Opening up a broad bean reveals the cotyledons, plumule and radicle.

The outer layers of the ovule become lignified forming a tough seed coat. The surrounding **ovary** develops into the **fruit** (Figure 4.28). The seed coat protects the embryo. In the final stage of its development the seed dehydrates, until only 5–10% of its mass is water.

Dormancy – waiting for the right conditions

Some seeds will germinate as soon as suitable conditions prevail but others are **dormant** and will not immediately germinate even in a favourable environment. Some of the ways that dormancy can be broken include:

1 exposure to an extended period of chilling
2 intense heat, for example in a natural fire
3 mechanical abrasion or microbial degradation of the seed coat
4 a minimum period of light (e.g. 10 hours)
5 chemical action in an animal's gut.

Q4.11 Suggest how each of the methods of breaking dormancy noted above could help ensure that the seed will germinate in favourable conditions.

How do plants disperse their seeds?

Different adaptations of seeds

Seeds come in all sizes and shapes, most of which are appropriate for wide dispersal. This helps offspring to avoid competing with their parent plant or with each other and it lets plants colonise new habitats. Remember that, of course, plants did not really design these structures. What we see are the successful results of evolution. The unsuccessful ones just don't exist any more – their line has become extinct.

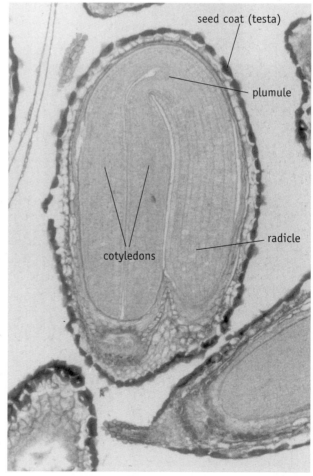

Figure 4.31 Embryo of shepherd's purse (*Capsella bursa-pastoris*).

There are four methods that seeds may rely on to carry them to their new location:

- wind dispersal
- animal dispersal
- water dispersal
- self dispersal.

Most species are adapted to take advantage of just one method.

As a seed develops, the surrounding ovary develops in parallel to form the fruit which has an important role in dispersal. Fruits can take many different forms as Figure 4.32 shows.

🔺 **Figure 4.32** Methods of seed dispersal used by different plants. Clockwise from top: poppy, cherry, goosegrass.

Q4.12 Look at this list of adaptive features and decide which method of dispersal is being exploited in each case: wind, water, animals or self-dispersal.
a extremely small and light seeds, e.g. orchids
b capsule-structured fruit ('pepperpot') supported on a long stem, e.g. poppy
c seeds with wings, parachutes or similar structures, e.g. sycamore, dandelion and willow herb
d dehiscence – explosive rupture of a seed pod, e.g. peas and laburnum
e hooked fruits, e.g. burdock and buttercup
f fleshy fruits, e.g. blackberry
g fibrous seed coats containing lots of air, e.g. waterlily and coconut.

Q4.13 Look at the photographs in Figure 4.32. For each plant identify the method of dispersal employed. Explain how the seed or fruit structure aids the dispersal.

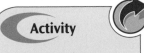

Activity

Activity 4.9 demonstrates how water can play an important role in the dispersal of some seeds. **A4.09S**

What happens when a seed germinates?

When conditions are suitable and any dormancy has been broken, the seed takes in water through a small pore in the seed coat, the micropyle. Cells in the embryo expand as they absorb water and the seed coat ruptures. Absorbing water triggers metabolic changes in the seed. Production of plant growth substances is switched on and these cause the secretion of enzymes that mobilise the stored food reserves. Maltase and amylase break the starch down into glucose which is converted to sucrose for transport to the radicle and plumule. Lipases break down stored lipids and proteases break down the proteins in the food store to give amino acids.

What can we do with starch from seeds?

Seeds are particularly useful in animal and human diets as a concentrated source of carbohydrate, lipids and protein. Although cereals are grown mostly for human food and animal feed, their carbohydrate polymers and oils also have major industrial uses.

Starch is easy to extract from plants because it is in granules which do not dissolve in water, but can be washed out. In wheat, the protein remaining is gluten, which is a rubbery mass. Its elastic properties are vital in bread-making but it has not yet found much industrial use.

We all eat starch, but it has many other uses. Starch is found in a wide range of products including adhesives, paints, textiles, plaster, insulating material and toiletries such as conditioners, mousses, sun screens and antiperspirants.

Activity

You can investigate the uses of starch for yourself in **Activity 4.10**. **A4.10S**

Thickening

When starch granules are heated in water they suddenly swell, absorb water and thicken the liquid. This is 'gelatinisation', and this thickening process is the basis of both custard and wallpaper paste!

Stiffening fabrics

The stiffening of cloth or paper by starch is known as 'sizing' and enormous amounts of starch are used in paper coatings and cloth treatment. A starch mixture applied to the surface is gelatinised and then cooled, allowing bonds to form between the starch molecules. The addition of water reverses the stiffening. This reversal is called 'plasticisation' (the material becomes flexible again) and water is the plasticiser. This reversible behaviour is very useful, since by the addition of a little water and heat, the shape can be changed many times.

Super absorbents

If starch is chemically cross-linked before it is gelatinised then particles are formed which can be dried. When rehydrated these particles can take up large amounts of water. Such cross-linked starch could soon be found as super absorbents in nappies and tampons.

Starch foam

The temperature at which starch gelatinises depends on the amount of water present. At water contents of less than 10% the gelatinisation occurs at much higher temperatures, above the boiling point of water. If the pressure is raised at the same time, the starch forms a plastic mass; if the pressure is suddenly released, for example when the seed coat ruptures during cooking, then steam forms and the starch 'puffs' into an expanded structure.

Puffed wheat breakfast cereals, corn snacks and starch-based foam packaging (Figure 4.33) are all made using this physical process, though corn snacks and starch-based foam are made in an extruder where a machine applies the pressure, rather than the seed coat. As hot starch leaves through a small exit hole the pressure that has built up inside is released, causing the starch to expand into foam as super-heated water turns to steam.

Starch-based packaging can now be used instead of polystyrene, polyethylene or other oil-based plastics. Plant-based plastics have also been developed which are made by fermentation of sugars from wheat, sugar beat, potatoes or agricultural waste. This means that supermarket food packaging such as a foam tray covered in plastic film could now be made from starch.

◀ **Figure 4.33** Foam packaging is commonly used to protect goods such as computers and telephones during transit from knocks and, to some extent, from large temperature fluctuations.

Activity

Have a go at popping corn yourself in **Activity 4.11** to demonstrate what happens to gelatinised starch when pressure is released. **A4.11S**

What can we do with vegetable oils?

Seeds are a rich source of oils which we regularly use in cooking. Oils are not only used for food – many other industrial uses have been developed.

Fuels

There is nothing new about using vegetable oils instead of petroleum-based products for motor vehicles. Castrol, the engine oil, was originally derived from the castor bean. Dr Rudolf Diesel developed the first diesel engine to run on vegetable oil, and Diesel's engine was demonstrated at the 1900 World Exhibition in Paris using peanut oil. Biodiesel produced today can be used in unmodified diesel engines alternating with petroleum diesel. It produces 100% less sulphur dioxide than diesel, and 80% less carbon dioxide when you take into account the carbon dioxide used by the plants grown to produce it.

In 1992 a trial was undertaken by the Reading Buses Company with three of their vehicles running on modified rapeseed oil. The results were very encouraging, but because the fuel was more expensive than petroleum diesel the trial was not extended. Such biodiesels are beginning to be used in some parts of the world.

Making soap

If oils and fats are boiled with an alkali such as sodium hydroxide, hydrolysis occurs to produce fatty acid salts and glycerol. The fatty acid salts are soaps, and the glycerol can be used in the manufacture of paint.

Q4.14 A simple soap will have the properties of the triglyceride fatty acids from which it was derived. If different triglyceride fatty acids are mixed, then harder or softer soaps can be formed depending on the melting points of the soaps. Will adding sodium palmitate to a soap of 'medium hardness' make it harder or softer? (Page 36 in Topic 1 shows the structure of palmitic acid.)

Plant products and sustainability

Will using starch packaging and other plant products make our use of resources more sustainable? **Sustainability** means that we can keep using the resources in the long term without harming the environment.

The use of oil-based plastics and fuels is not sustainable for several reasons:
- burning oil-based fossil fuels releases carbon dioxide into the atmosphere contributing to global warming (see later in this topic, pages 207–208)
- oil reserves will eventually run out
- plastics generate non-biodegradable waste, creating major waste disposal problems.

The use of plant-based products should help reduce these problems. Although burning fuel made from vegetable oil also produces carbon dioxide, carbon dioxide has been removed from the atmosphere relatively recently when the crop that produced the oil was grown.

However, sustainability is not guaranteed by simply switching to products made from plants that were alive very recently rather than fossil plants. We also need to consider the energy used and any pollution created during the production and transport of the product.

Exchanging plastic bags for paper ones in supermarkets might seem like a good idea, but if the bag gets wet the shopping ends up on the floor as the bag disintegrates. A plant-based plastic bag that remains intact in the wet, is light and will also biodegrade would be a sustainable alternative.

Does biodegradable packaging solve our waste problem? Most bacteria involved in degradation are aerobic, so what happens if the packaging is buried deep in a landfill site? The answer is remarkably little – newspapers have been retrieved from deep landfill after 10 years and have still been readable! Even a frankfurter had not degraded very much. Closer to the surface, packaging will degrade, though sometimes methane, which contributes to global warming, is generated in such environments.

4.4 Manipulating plant biology for human ends

As far back as the Stone Age, farmers were growing and modifying plants. Crops were first cultivated some 10 000 years ago in the Middle East, the Americas and the Orient. Some of the wild ancestors of today's crops can still be found although they are very different from their domesticated descendants. For example, wild potatoes are poisonous, and wild sugar cane produces very little sugar.

How were traditional crops produced?

Selecting the best plants

Throughout history, humans have selected particular plants for cultivation as crops, because they may be edible, produce good fibres, contain useful chemicals or just look ornamental. For centuries farmers have picked out the hardiest and most prolific plants from their crops and have saved the seeds from these plants for sowing the following year. In this way crops have steadily improved, the best plants from one season being used as parents for the next. Individual plants with poor yield, bad flavour or prone to disease have not been chosen. This process is called artificial selection. It selects alleles for characteristics which are agriculturally valuable. For example, 50 years of artificial selection have doubled the oil content of the maize varieties that produce cooking oil.

Q4.15 What other characteristics might be selected as valuable features for crop plants?

Where do these special plants come from?

Quite often a new form of a plant grows by chance as a result of changes to the DNA during replication or meiosis. Chemical or physical **mutagens**, environmental factors such as gamma radiation that increase the rate of mutation, can be used to bring about more mutations and so increase the chance of producing plants with new characteristics. Many of these new plant forms die or are not fertile but a few may be useful for breeding.

▲ **Figure 4.34** Test beds at the Cambridge Plant Breeding Institute where new forms of a plant with the desired characteristic are identified for future breeding.

Today a plant breeder will identify individual plants with desired characteristics. These may then be self-pollinated if the plant has both male and female parts on the same individual. The seeds are grown in long rows (Figure 4.34) and the best offspring are selected and again self-pollinated. If this process is repeated several times the genetic variation is reduced and the plant becomes more homozygous. In this way new varieties are produced; each pure-bred line from an individual plant may give rise to a new variety (Figure 4.35).

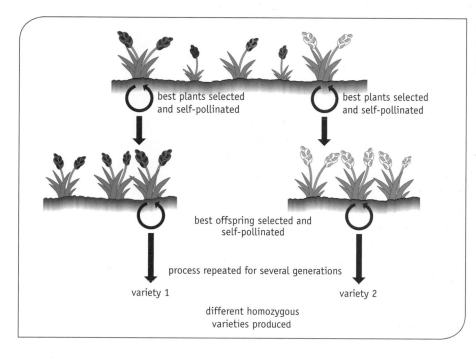

Figure 4.35 New plant varieties are often produced by selecting and self-pollinating plants with desired characteristics.

Inbreeding problems

However, if the plant does not normally self-pollinate, this inbreeding can result in a reduction in vigour with a loss of size, yield and fertility. This is called **inbreeding depression**. It occurs because crossing closely related individuals is more likely to result in offspring with two copies of harmful recessive alleles. However, cross-breeding two inbred lines typically produces a hybrid that is more vigorous than either of the parents, a phenomenon known as **hybrid vigour**. The offspring are heterozygous at many loci. Many seeds used by farmers and gardeners are F_1 hybrids, that is, first generation hybrids produced by crossing two pure-bred lines (Figures 4.36 and 4.37).

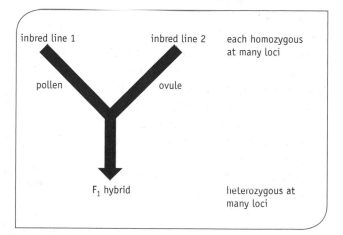

Figure 4.36 A cross between two inbred lines produces an F_1 hybrid.

Crossing different varieties

On many occasions useful traits have been combined by **hybridising** two different varieties or indeed two closely related species. At least two separate such **interspecific** crosses, thousands of years ago, provided the ancestors for the modern wheat we use for bread. More recently, during the last hundred years, plant breeders have hybridised wheat with rye to produce triticale. Triticale combines the high quality and yield of wheat with the tolerance to weather extremes and the resistance to fungal infection of rye. It is used widely for animal feed and is particularly important in eastern Europe.

Plant breeding is a slow process, and from identifying a potential new variety to commercial production takes many years. But suppose we were able to change precisely just one or two instructions in the genetic code. Then we could be much more specific in the properties of the new plant, and get a much quicker result than by conventional breeding. We could even take an instruction from one plant species, or even an animal or bacterium, and insert it into our crop species, where that instruction was previously not present in any possible parent to the plant.

Until the 1980s this was not possible. Then came genetic manipulation, also known as genetic engineering and genetic modification.

🔺 **Figure 4.37** A maize hybrid (centre) produced by crossing two inbred lines (left and right) is much more vigorous than either parent.

What is genetic manipulation?

Genetically modified (GM) crops are produced not by conventional breeding techniques, but by a laboratory process whereby genetic engineers introduce new genes with alleles for a desired characteristic into a plant's DNA. The resulting transgenic plants can be produced in a timescale of months rather than the years required for traditional methods (Figure 4.38).

How are the new genes inserted?

Foreign genes are inserted into plant cells. This can be achieved using a range of methods:

- A bacterium which infects many species of plant, such as the soil-inhabiting bacterium *Agrobacterium tumefaciens*, can be used. The bacterium contains small circular plasmids of DNA as well as a single much larger, circular chromosome. When the bacteria invade plant cells, genes from the plasmid DNA become incorporated into the chromosomes of the plant cells. Scientists have developed a technique using this natural system. They insert the desired genes into a plasmid which then 'carries' these genes into the plant DNA.

🔺 **Figure 4.38** Transgenic Bt tobacco has been modified by inserting a bacterial gene which allows the plant to produce its own insecticide, protecting the plant from damaging insect pests; see page 178.

Did you know? Genetically engineering bacteria

The first successes with the technology of genetic engineering were not with plants but with bacteria. Bacteria contain simple DNA structures, plasmids, which can be transferred from one cell to another. Using restriction enzymes that act like scissors, the circular plasmid can be cut, and using another set of enzymes a piece of DNA from another species can be inserted. The DNA is then replaced in the bacteria.

Figure 4.39 shows how this process is used for producing the very valuable human protein, insulin.

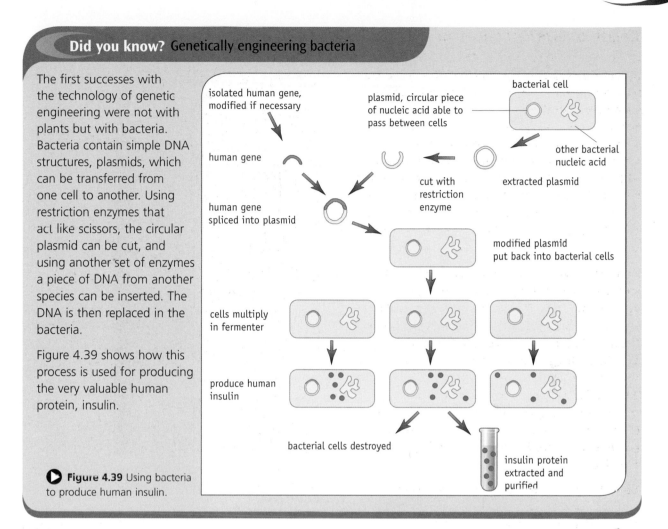

▶ **Figure 4.39** Using bacteria to produce human insulin.

- Minute pellets which are covered with DNA carrying the desired genes are shot into plant cells using a particle gun.
- **Viruses** are sometimes used. They infect cells by inserting their DNA or RNA (depending on the type of virus). They can be used to transfer the new genes into the cell.

Gene insertion is never 100% successful. Scientists therefore need a method of screening to find out which plant cells actually have the new gene. This has generally been done by incorporating a gene for antibiotic resistance, often called a **marker gene**, along with the new desired gene. The plant cells are then incubated with the antibiotic, which kills off any unsuccessful ones which have not taken up the new genes. The only cells which survive are those which have successfully incorporated the new genes and are resistant.

Growing the new plant

The genetically modified plant cells can then be 'cultured' in agar with nutrients and plant growth substances, to produce new plants. This **micropropagation** can multiply a single cell to form a callus (a mass of plant cells) which then differentiates to form plantlets (tiny plants) (Figures 4.40 and 4.41) and finally novel plants.

Activity

Use the interactive tutorial in **Activity 4.12** to produce your own summary explaining how plants can be genetically modified. **A4.12S**

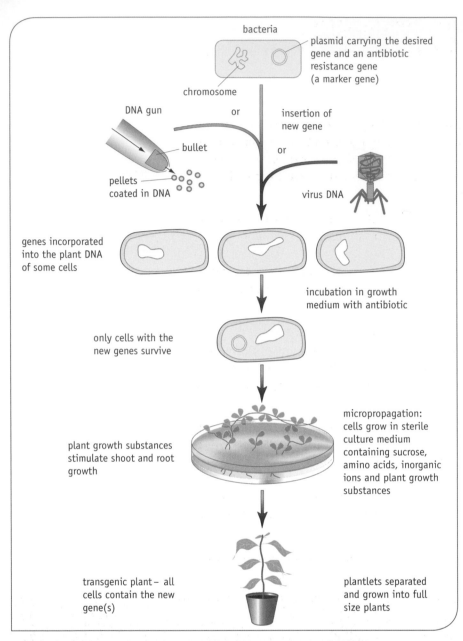

bacteria

plasmid carrying the desired gene and an antibiotic resistance gene (a marker gene)

chromosome

DNA gun

or

insertion of new gene

bullet

or

pellets coated in DNA

virus DNA

genes incorporated into the plant DNA of some cells

incubation in growth medium with antibiotic

only cells with the new genes survive

microproagation: cells grow in sterile culture medium containing sucrose, amino acids, inorganic ions and plant growth substances

plant growth substances stimulate shoot and root growth

transgenic plant – all cells contain the new gene(s)

plantlets separated and grown into full size plants

◀ **Figure 4.40** Production of genetically modified plants.

◀ **Figure 4.41** A callus of parenchyma cells is treated with growth substances that enable cells to differentiate into shoots and root tissue.

The potential of genetic engineering

A wide range of genetically modified crops have already been produced. Some examples are described here.

Tougher tomatoes

The first case of the genetic engineering of a plant for food use did not involve introducing a gene from another plant or even from an animal or a bacterium. Instead, an artificial gene was made and inserted into a tomato. This gene had the effect of preventing the modified tomato from making an enzyme which causes the fruit to soften.

Normal tomatoes make an enzyme called polygalacturonase – PG for short. PG breaks down some of the compounds in the cell walls of a ripe tomato. It therefore causes tomato fruits to go soft. In a wild tomato this is useful as birds are attracted, eat the fruit and so disperse the tomato's seeds. However PG is a nuisance for tomato growers as soft tomatoes are more likely to get damaged before they can be sold.

The first 15 bases of the coding region of the PG gene are ATGGTTATCCAAAGG. The gene that was made in the laboratory and inserted into the tomato had a base sequence that began TACCAATAGGTTTCC. What do you notice about these two DNA sequences? You should spot that they are complementary.

Now think what happens when both the PG gene and the introduced artificial gene make messenger RNA. (You might like to work out the first 15 bases of these two mRNAs.) What do you notice about these two mRNA sequences? In the tomato cells what happens is that these two sorts of mRNAs with their complementary sequences base pair with each other in the cytoplasm. Indeed, they stick together so firmly (as a result of hydrogen bonding) that neither mRNA makes much protein.

Q4.16 Draw a diagram to show how the inserted gene prevents the cell making the enzyme PG.

Measurements show that PG activity in these GM (genetically modified) tomatoes is only about 1% of normal. As a result, the tomatoes stay firmer for longer and the grower saves money. The tomato was patented and trademarked as the Flavr Savr (Figure 4.42).

🔺 **Figure 4.42** Tomato paste made with genetically modified Flav Savr tomatoes is no longer available in UK shops.

Roundup Ready crops

Farmers have always battled with weeds. Fast-growing and rapidly spreading, these unwanted plants compete with crops for water, nutrients, light and space. If they are left unchecked, weeds can decimate a farmer's crops and destroy his or her livelihood. The age-old and traditional method of control was 'weeding' but on most farms in developed countries this has now been replaced by herbicide sprays. In Europe and the US over 90% of arable land is now treated with these chemical weedkillers.

Glyphosate is known as a 'broad-spectrum' herbicide because it kills nearly anything green. It is the main component of 'Roundup', the world's top-selling herbicide, manufactured by Monsanto. Because Roundup is toxic to crops as well as to weeds, it used to be sprayed only onto fallow (unplanted) fields and the borders of planted areas. Here it rapidly breaks down to harmless products leaving the soil safe and weed-free. Once the crop is growing, 'narrow-spectrum' herbicides must be used: these leave the crop undamaged but target specific weeds.

Plants such as cotton, soybean and tobacco (Figure 4.38) have been genetically modified so that they can survive direct applications of Roundup. The resistance comes from a bacterial gene inserted into the DNA of the plant. 'Roundup Ready' seeds for GM plants resistant to Roundup are also produced by Monsanto. This means that farmers, instead of relying on a yearly schedule of different herbicides, can simply rely on Roundup. The entire crop can be doused with one herbicide which kills all the weeds while leaving the GM crop unharmed. Many see this as an agricultural breakthrough: the costs of weed control are less for the farmer; the profits for Monsanto are higher.

Bt corn and the corn borer problem

Maize (corn) plants are frequently plagued by a damaging caterpillar (Figure 4.43). The European corn borer larvae feed on the young leaves of maize plants and then bore into the stems. Here they cause damage and blockages preventing the normal transport of water, minerals and sugars. As a result the infested maize plants grow feebly and only produce puny cobs for harvesting. In extreme cases the invaded stalks break and the crop is totally wasted.

Agriculturalists estimate that, worldwide, the European corn borer is responsible for an average yield loss of about 7% of the maize crop each year. This is equivalent to 40 million tonnes of maize, worth about US$2 billion.

For years farmers have tackled the problem using both biological control and chemical insecticides. The biological control involves spraying the crop with *Bacillus thuringiensis*, a species of soil bacterium which naturally produces a toxic protein, lethal to the corn borer but harmless to vertebrates. These methods have little success, largely because the corn borer caterpillars are hidden within the stems, well protected from both chemical pesticides and bacterial sprays.

⬤ **Figure 4.43** The corn borer within the maize stem is protected from any insecticide sprayed on the plant.

New hopes for controlling the pests have come with the development of transgenic Bt corn. These maize plants have been genetically modified by inserting the bacterial gene for the Bt (*Bacillus thuringiensis*) toxin. Because every cell in these plants makes the Bt toxin the corn borers cannot avoid being poisoned. As the caterpillars feed on the plant cells, their intestinal walls become damaged and they soon die. Early trials of Bt corn indicated an increased yield of 5–10% and by 2001 about half of the US maize fields were sown with the new transgenic variety.

Engineering metabolic pathways

All of these examples require the introduction (or silencing) of a single gene, which gives the new plant advantages over traditional varieties. By now, several genes can be inserted at once, which allows the modification of whole chemical pathways in the plant. Manipulation of the lignin pathway is producing trees whose cell walls are more easily turned to paper pulp. This process where whole pathways are re-engineered is known as 'metabolic engineering'.

Work is underway to produce crops with, among other characteristics, increased yield, improved flavour, better processing characteristics and higher food quality. This includes the development of crops with high levels of vitamins or antioxidants for health protection. For example, 'golden' rice contains high levels of vitamin A to reduce the incidence of certain sorts of anaemia and blindness. In the future plants may produce vaccines against human or animal disease and even precursors for plastics and silks.

Checkpoint

4.4 Explain how the genetic modification of plants is similar to but distinct from conventional breeding.

Concerns about genetic engineering

So far so good. There are potential benefits of GM technology but as with the advent of any new major technology (such as nuclear power) we also need to examine carefully the risks associated with its use.

Health

The main health concerns are:
- transfer of antibiotic resistance genes to microbes
- formation of harmful products by new genes.

Antibiotic resistance

GM foods often contain not only the particular gene (for pesticide resistance, a higher yield, or whatever) that has been inserted into the crop but also a marker gene to select for the new plants. These, as discussed above (page 175), are sometimes antibiotic-resistant genes. For example, Bt corn contains a marker gene which gives resistance to ampicillin and other penicillin-type antibiotics. When eaten the gene could potentially be transferred to microbes in the gut which could build up resistance to certain antibiotics used in medical treatments.

> ### Did you know? The problem of antibiotic-resistant bacteria
>
> Already the overuse of antibiotics, both in medicine and in animal feeds, has led to disease-causing bacteria becoming resistant to almost all the antibiotics we have. Many antibiotics which were once commonly used are now ineffective. Pharmaceutical companies are having, at great expense, to develop new compounds to combat new, resistant strains of pathogen. For example methicillin and flucloxacillin have largely replaced the old penicillin.
>
> Outbreaks of infection caused by antibiotic-resistant bacteria periodically hit the headlines. In 1999, Kettering Hospital had to close a ward and spend nearly half a million pounds disinfecting it after an outbreak of methicillin-resistant *Staphylococcus*
>
> *aureus* (MRSA). Stories like this are becoming more common. *Staphylococcus aureus*, or 'Staph' as it is often called, is widespread on human skin and elsewhere. Although normally harmless, if it enters the wounds of weak post-operative people it can cause blood poisoning and other life-threatening conditions. The treatment is with methicillin but this cannot help if the patient is infected with MRSA bacteria.
>
> Patients are also at risk from other multidrug-resistant organisms, for example PRSP (penicillin-resistant *Streptococcus pneumoniae*) which causes a type of pneumonia. Antibiotic resistance is already a medical problem. Could the introduction of Bt corn make it worse?

In 1996 the EU banned Ciba-Geigy's Bt corn on the grounds that it also contained a potentially hazardous gene for ampicillin resistance. (This outright ban has since been replaced by a requirement for specific labelling.) In 1999 the British Medical Association urged a ban on all use of antibiotic-resistance genes in GM foods, stating that 'the risk to human health from antibiotic resistance developing in microorganisms is one of the major public health threats that will be faced in the twenty-first century'.

Since antibiotic-marker genes are there only to assist the scientists in developing the new crop, and do not benefit anyone once the crop is on the market, it seems unnecessary to leave them in the products. Routes for their removal are already available.

Harmful products from new genes

Could the substances made by the new genes in GM crops be harmful in any way? It is difficult to guarantee that any activity in life is risk free – as we saw back at the beginning of this AS course (pages 20–22). So far there are no reported cases of ill health resulting from the consumption of GM crops. However, biochemical changes to oils, proteins and other plant substances might conceivably result in toxic compounds or new allergens. The law requires that these possibilities have to be investigated thoroughly before any GM food is introduced.

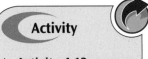

Activity

In **Activity 4.13** you can consider the benefits and risks of genetically modified foods. **A4.13S**

Choice

People are free to eat whatever they like provided it is safe. We do not require people to eat particular foods, and, for personal or religious reasons, many safe foods are not consumed by certain sectors of the population. If only a few genes are changed in a GM crop, and the product is safe to eat, shouldn't people be allowed to choose whether or not they would like to eat food made from it?

Environmental issues

The main environmental concerns about GM plants are:
- transfer of genes to non-GM plants
- increased chemical use in crops.

Transfer of genes to non-GM plants

Cross-pollination, transfer of pollen from one plant to another, can occur over quite long distances by wind or insects. Some of the crops we grow are related to wild plants living nearby and can cross with them (for example, oil seed rape is related to wild mustard). Certainly, GM crops can cross with conventional crops of the same species growing in nearby fields. This means that genes introduced into a GM crop will inevitably spread to conventional crops – and this has been shown to occur. They may also be transferred to wild plants. Once they have 'escaped' such genes cannot be recalled.

Creation of superweeds

Most of our crops cannot survive in the wild – they are not fit enough. Wild species would soon overrun our fields, if crops were not given a helping hand by pesticides, fertilisers, herbicides and mechanical weeding. Therefore any escaped transgenic crop plants might be presumed to disappear. But suppose the introduced gene makes the crop or any plants it is transferred to much fitter, for example, capable of surviving drought, frost or insect attack. Then it could have an advantage, even over its wild relatives. Then we might have 'superweeds'.

We already have examples of 'superweeds' – without genetic engineering – due to the introduction of plants, such as rhododendrons and Japanese knotweed, from other countries. Some such plants have proved very well suited to survive in our environments and have caused considerable ecological havoc.

Increased herbicide use

In the US, wild mustard that is resistant to Roundup herbicide has invaded fields of Roundup Ready oil seed rape, and resistant mare's-tail has been found amongst Roundup Ready soybean crops. Resistant rye grass has been reported in Australia and resistant goose grass in Malaysia. Organisations that campaign against GM technology such as Greenpeace and Friends of the Earth argue that far from reducing the extent of pesticide use, GM crops will end up causing us to use even more chemicals to control resistant weeds and insects than we would otherwise have done.

One solution to the problem of GM superweeds would be to ensure that the outcrosses (crosses between GM plants and other plants of a different variety) are not fertile and cannot proliferate. An alternative is the development of technology whereby the pollen does not contain the modified gene so it cannot spread. Both approaches are being investigated, but are some way off.

Who owns these new plants?

Even before genetic engineering was possible, plant breeders could protect a new variety under the Plant Breeders' Rights Act. They 'owned' the seed, but many farmers kept back seed from each harvest for the next year to avoid buying new ones. The advantage to the farmers was that they saved money. One disadvantage is that the saved seed is sometimes not as good as the original (for example, if the original was an F_1 hybrid – see page 173). Biotechnology companies now patent the new technology used in gene modification, the introduced genes and, indeed, new varieties. Biotechnologists claim that they have spent time and money 'inventing' the new crop and deserve patents just like any other inventor.

Developing countries are unlikely to be able to afford to pay for licences to use the new technology so that they can produce GM crops with characteristics suited to local conditions, such as drought resistance or salt tolerance. Ownership of this powerful new technology by a handful of huge multinational corporations can threaten less developed societies.

Farmers with enough capital may be able to benefit from GM technology, producing high-yield crops which they can then sell at competitive prices. Most of these farmers live in relatively affluent countries which already have food surpluses. In contrast small farmers, living in less developed areas where there are often food shortages, cannot afford to invest in the pricey seeds and necessary agrochemical backup. Moreover, when they come to sell their traditional crops they may find that they have to compete with the cheaper products from GM varieties. As a result, they may well get even poorer, fall into debt and be forced to sell off their small plots of land. It is rarely easy to predict the social consequences of scientific developments.

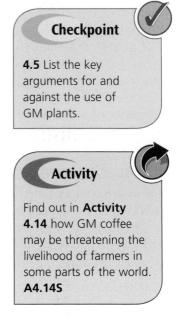

Checkpoint

4.5 List the key arguments for and against the use of GM plants.

Activity

Find out in **Activity 4.14** how GM coffee may be threatening the livelihood of farmers in some parts of the world. **A4.14S**

4.5 Coping with climate change

One of the biggest challenges facing plants, and for that matter all other organisms, is climate change. If climate change is caused by global warming, and we do appear to be in a period of global warming, we will probably see significant changes in the flora (plants) and fauna (animals) of Britain as well as other parts of the world.

Some animals can migrate to avoid the changing conditions. Plants, even though they are rooted to one spot and cannot simply move, will change their distribution over time. The changes that are already being attributed to climate change can be divided into two categories:
- changing distribution of species
- altered development and life cycles.

Changing distribution of species

Changing communities and alien invaders

A **community** is a group of species found in the same place at the same time. Within any group of species, some will cope with change whilst others will fare less well. Climate change will therefore cause the balance between species in the community to shift. Some species may benefit from the new conditions and become **dominant**, whilst others may be lost from the community altogether due to competition with existing or invading species. If they are mobile or have good seed dispersal they may migrate to more favourable conditions. In other words, the distribution of some species will probably change. For example, plants that currently reach their northern limit of distribution in England or southern Scotland, such as the ground thistle, will begin to expand northwards. Others, such as heath bedstraw (Figure 4.44), will retreat north from the southern limits of their ranges in Britain. Future communities may well be quite different to those present today.

Q4.17 Suggest why heath bedstraw distribution will move further north with climate change.

Activity

Activity 4.15 allows you to find out about coral bleaching linked to rising sea temperatures. **A4.15S**

◀ **Figure 4.44** Heath bedstraw distribution will move north if conditions in the south become warmer.

A study of 35 non-migratory European butterfly species found that the ranges of 63% of them have shifted northwards by between 35 and 240 km during the twentieth century. Figure 4.45 shows how the speckled wood butterfly (Figure 4.46) has moved north in Britain. This shift is a result of more successful colonisation at the northern edge of their range and an increase in extinctions at the southern edge. Only 3% of the species studied had shifted further south. The changing distribution may be a direct response to rising temperature, or the result of a shift in distribution of the plants they feed on.

With global warming, higher temperatures early in the year could extend the growing season and crops such as olives and citrus that are sensitive to frosts could be grown much further north.

Q4.18 In mountainous regions, how might the ranges of plants and animals shift with warmer conditions?

A particular problem for some native communities may be the invasion of exotic or alien animal or plant species from other regions of the world. Species from Mediterranean regions might take advantage of the warmer conditions and invade southern England, pushing out the current inhabitants. Some exotic species such as scorpions already survive in local areas in Britain and could spread rapidly given a suitable climate.

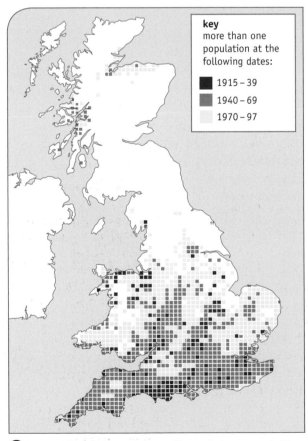

key
more than one
population at the
following dates:

■ 1915 – 39
■ 1940 – 69
□ 1970 – 97

▲ **Figure 4.45** The speckled wood butterfly has extended its range north. *Source: Nature.*

▲ **Figure 4.46** The speckled wood butterfly.

Pests and diseases may also spread to new areas and act to reduce crop yields. Witchweed is a **parasitic** weed that infects cereal crops including maize and sorghum in Africa (Figure 4.47). It already causes huge amounts of damage to subsistence farms, but requires an air temperature of at least 28 °C to grow effectively. Unlike most weeds, which merely compete with crops, a parasitic plant like witchweed taps directly into its host plant and absorbs nutrients and moisture. As a result the host plant is less productive, destroying 30–100% of the crop. If global temperatures increase, it may spread to cereal crops in new areas and cause even more damage.

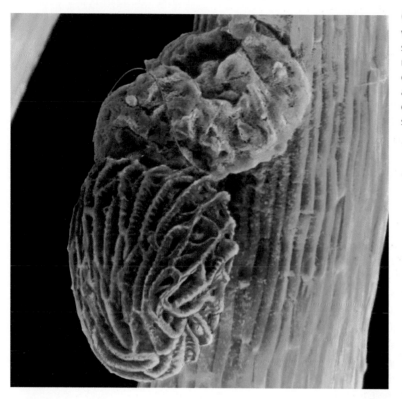

Figure 4.47 The parasitic witchweed already causes serious damage in Africa and may become more widespread if conditions become warmer. Here a witchweed germinates and establishes itself on the host crop's stem. Magnification ×115.

In Britain, fruit crops could be threatened by the easier spread of fungal diseases in more humid conditions.

Some invertebrate pests are also likely to change their distribution. For example, the nematode *Longidorus caespiticola*, which feeds on grass, is currently restricted to England, Wales and south-east Scotland. A 1 °C rise in temperature could extend its range to cover the whole of Scotland. The brassica pod midge *Dasineura brassicae*, a major pest of oil seed rape, is also expected to move further north. Not only is it likely that the distribution of such pests will change, but the warmer temperatures may allow them to produce more new generations in a single year, increasing their effect on crops. However, it is hard to predict what the impact will really be due to interactions with other predators and any change in the life cycle of the host plant.

It is not only changing temperature that affects species distribution. Changing rainfall patterns, soil moisture, winds and rising sea levels are all likely to influence species and communities.

Weblink

Visit the Department for Environment, Food and Rural Affairs website where you can read reports on climate change and agriculture in the UK.

Extension

In **Extension 4.1** you can find out about what might have caused extinction of the dinosaurs. **X4.01S**

Altered development

Faster photosynthesis – faster growth

Scientists have established that a few degrees of warming will lead to an increase in temperate crop yields, but higher temperature rises will have a negative effect on crop yields. As you probably already know, plant growth depends on photosynthesis, which we can summarise in the simple equation:

$$\text{carbon dioxide} + \text{water} \xrightarrow{\text{light}} \text{glucose} + \text{oxygen}$$
$$6CO_2 \quad + 6H_2O \longrightarrow C_6H_{12}O_6 + \quad 6O_2$$

In turn, the rate of photosynthesis is determined by a number of factors including both carbon dioxide concentration and temperature. In cooler climates where photosynthesis is temperature limited, a rise in temperature will result in faster photosynthesis. But the situation is a little more complicated because above an optimum temperature (which varies for different species) plant enzymes work less efficiently. Figure 4.48 shows how the rate of an enzyme-controlled reaction is affected by temperature.

Key biological principle: The effect of temperature on enzyme activity

Temperature affects the rate of an enzyme-catalysed reaction, as Figure 4.48 shows. At low temperatures the reaction is very slow. This is because the enzyme and substrate molecules move slowly and don't collide very often. As the temperature increases there are more collisions so the substrate binds with the enzyme's active site more frequently thus increasing the rate of reaction. The temperature at which the rate of reaction is highest is called the **optimum temperature** (Figure 4.48). If the temperature continues to rise the enzyme molecule vibrates more and bonds that hold it in its precise three-dimensional shape break. The substrate no longer fits easily into the active site; this slows the reaction. Eventually the shape of the active site is lost and the enzyme-substrate complex no longer forms; no reaction can occur. The enzyme is said to be **denatured**.

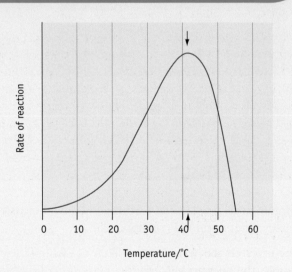

◢ **Figure 4.48** The effect of temperature on the rate of reaction of an enzyme-catalysed reaction. The optimum temperature is indicated by an arrow.

Q4.19 **a** Look at Figure 4.49 and decide on the optimum temperature for wheat photosynthesis under high light conditions.
b Most animal species are much more mobile than plants and some migrate between regions with widely different temperatures. Animals such as migratory fish already show the ability to cope with a range of temperatures. They often synthesise slightly different forms of the same enzyme, termed isoenzymes, as they move from one region to another. How might having different forms of the same enzyme help them cope with changing conditions?

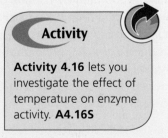

Activity

Activity 4.16 lets you investigate the effect of temperature on enzyme activity. **A4.16S**

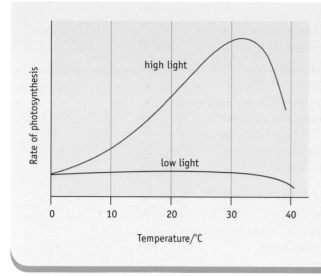

Increases in temperature result in higher rates of photosynthesis when no other factors are limiting. Above an optimum temperature, where the rate of photosynthesis is at a maximum, rate declines. Under natural conditions the optimum is rarely achieved because CO_2 and light are limiting.

Figure 4.49 The response of photosynthesis in wheat to increasing temperature at two light intensities.

A complex picture

In today's tropics, many crops are at the limit of their temperature tolerance, and arid conditions dominate. Even a small rise in temperature would be likely to decrease yield by the direct effect of the higher temperature along with the resulting fall in soil moisture.

Predictions about the ecological and agricultural consequences of global warming are difficult to make with great confidence, as there are so many interacting factors. If the concentration of carbon dioxide in the atmosphere rises, the rate of photosynthesis in many plant species may also increase, irrespective of any rise in temperature resulting from the rising carbon dioxide levels. This is because carbon dioxide is a **limiting factor** for photosynthesis. The graph in Figure 4.50 illustrates another point about limiting factors – the rate of photosynthesis would not increase indefinitely with rising carbon dioxide, because other factors, such as light intensity, water or nutrients would start to limit photosynthesis.

Overall, it is probable that crop production in cooler temperate regions will benefit from climate change whereas warmer tropical regions may suffer from poorer yields. Given that the poorest countries in the world are in the tropics, global warming may increase global inequality.

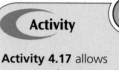

Activity

Activity 4.17 allows you to analyse research data and design an experiment to investigate the effect of changing environmental conditions on plant growth. **A4.17S**

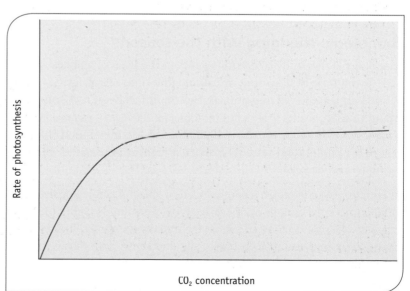

Figure 4.50 The response of potato leaves to increased carbon dioxide concentration.

Disrupted development and life cycles

Animals are likely to be affected if temperature acts as an environmental cue or trigger for their development or behaviour. In the case of salmonid fish (salmon, trout and sea trout), it is their spawning, hatching and growth rates that are most likely to be disrupted. Recent research has shown that brown trout stop growing in late summer when the river temperatures have warmed up. At present, this is not a problem because they have already completed most of their growth by this stage in the year. However, if river temperatures were to warm up earlier, they might stop growing earlier, resulting in underweight fish with a reduced chance of surviving the winter.

The egg incubation temperature of some reptiles determines the sex of the offspring. In leatherback turtles, higher temperatures in the nest result in females (Figure 4.51). Usually, there are sufficient nests at or below the critical temperature for the sex ratio to remain reasonably close to 50:50.

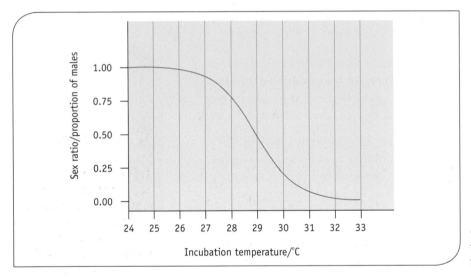

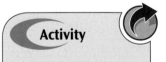 **Figure 4.51** In leatherback turtles, temperature determines the sex of the hatchlings.

Q4.20 What would be the implications of warmer temperatures for leatherback turtles?

Phenology: changing with the seasons

Phenology is the study of natural events in the lives of animals and plants, such as the time of flowering or fruiting, the time of egg laying or hatching, the first appearance of migrants and so on. It has been recognised that the timing of such events is a useful biological indicator of global climate change. Many long-term records exist that allow us to see how much change has been experienced in natural systems, particularly those associated with the onset of spring (Figure 4.52).

Of course, evidence from a single location about a single species is not sufficient to substantiate claims that global warming is taking place. However, data sets exist for a range of events from many locations. Trends within them can be identified and tested statistically, and these do indicate that spring is getting earlier for many species.

Activity

Activity 4.18 allows you to investigate the effect of temperature on the hatching success of brine shrimps. **A4.18S**

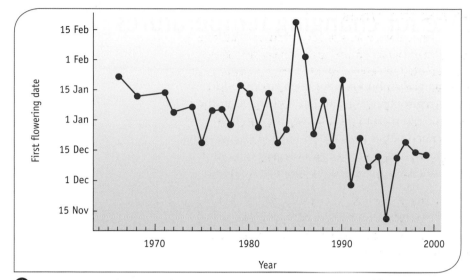

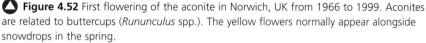

● **Figure 4.52** First flowering of the aconite in Norwich, UK from 1966 to 1999. Aconites are related to buttercups (*Rununculus* spp.). The yellow flowers normally appear alongside snowdrops in the spring.

For example, Richard Fitter, one of Britain's leading naturalists, recorded the first date on which over 500 plant species flowered in each year from 1954 to 2000 near his home in Oxfordshire. Professor Alistair Fitter, his son, has analysed these data. He has found that nearly one in six of the species are now flowering significantly earlier than they did before the 1990s. For example, white dead-nettle flowered 55 days earlier in the 1990s compared with the period 1954–90, and can now often be seen flowering throughout the winter. However, not all species responded as dramatically as this, and a few flowered later.

The UK Phenology Network in Britain aim to build on these existing data sets. Leafing, flowering, summer bird arrival, bird activity and the behaviour of insects, butterflies and amphibians are all being monitored.

Making the most of food supplied

For many species, the hatching of eggs or the emergence of adults is synchronised with periods of maximum food availability. For example, the eggs of many marine worms are laid so that hatching coincides with high levels of microscopic plants (**phytoplankton**) on which the worm larvae feed. The problem is that the worms lay their eggs in response to day length (**photoperiod**), whereas phytoplankton grow in response to temperature. If global temperatures rise, the peak in phytoplankton will occur earlier in the spring, but the worms will still lay their eggs at the same time because the photoperiod is unchanged. The resulting mismatch between hatching time and peak food availability could drastically reduce the survival rates of the worm larvae. This lack of synchrony is also observed with UK birds such as great tits and one of their food sources, winter moth caterpillars.

Checkpoint

4.6 Write a short summary that explains how rising temperatures, changing rainfall patterns and changes in seasonal cycles can affect plants and animals.

Activity

Activity 4.19 involves you in analysing an online dataset for the emergence dates of some butterflies and moths, and visiting the UK Phenology Network interactive garden to investigate some seasonal changes affected by rising temperatures. **A4.19S**

4.6 Evidence for changing temperatures

Observed changes in plant and animal distribution have been attributed to rising temperatures. Is the climate really changing, and has it changed in the past? How can we tell? Older people often claim that the weather in their childhood was different – usually much better than today! Such evidence from personal memory is termed anecdotal, but is often unreliable and can only go back less than 100 years. We cannot rely on anecdotal reports; we need reliable scientific evidence.

We seem at present to be in a period of global warming. This view is supported by evidence from a range of sources including:

- temperature records
- pollen in peat bogs
- dendrochronology (tree-ring studies).

Temperature records over long periods

Long sequences of temperature records exist for a number of places, for example central England from 1659 to the present and Toronto, Canada from 1780 to the present. Old datasets are very important in the study of climate change, even though they may have been collected with equipment which may not be as accurate as those used today.

Q4.21 Look at Figure 4.53. The graph shows a long dataset of temperature measurements for central England. Do the data show an increase in temperature? To help you decide, you could place a piece of tracing paper over the graph and draw a straight best fit line and a separate curved best fit line. These would help you to pick out any trends in the data.

Activity

In **Activity 4.20** you analyse some long datasets for yourself.
A4.20S

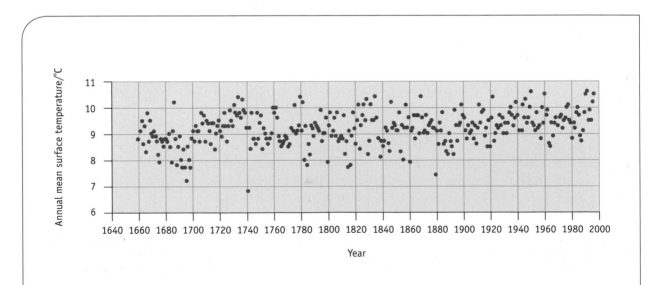

▲ **Figure 4.53** Is the temperature rising? *Source: Open University.*

Studying peat bogs

Records of temperatures measured with a thermometer only take us back two or three centuries at the most. One way of finding out about the climate at least back to the last Ice Age (which ended in Britain around 12 000 years ago) is to study plant and insect remains preserved in peat (Figure 4.54). Peat is an accumulation of poorly decayed organic matter, mainly the remains of plants long dead. There are still extensive peat bogs in Ireland and some upland areas of Britain.

When plant material dies it normally decays. However, in the anaerobic and often acidic conditions of a peat bog, the decay rate is slowed or stopped altogether.

Q4.22 Why do anaerobic and acidic soil conditions slow the decay rate?

Pollen from the past

Many objects are found preserved in peat, such as Viking ploughshares, prehistoric tree trunks and even human bodies. **Pollen grains** are particularly well preserved and can be used to determine climate conditions in the past. Pollen from peat is useful for reconstructing past climates because:

* Plants produce pollen in vast amounts. Throughout the spring and summer, countless millions of pollen grains fall from the air onto the ground, including the surface of peat bogs, as 'pollen rain'.
* Pollen grains have a tough outer layer which is very resistant to decay.
* Each species of plant has a distinctive type of pollen allowing us to identify the plant species from which it came.
* Peat forms in layers: the deeper the layer, the older the peat. Carbon-14 dating allows the age of a particular peat layer to be established.
* Each species of plant has a particular set of ecological conditions in which it flourishes best. If we find pollen from a species that favours warmer conditions, we can infer that the peat was laid down when the climate was warmer.

▲ **Figure 4.54** A scientist takes a core from a peat bog. The type of pollen found in each layer of the core provides information about the conditions when the peat was deposited.

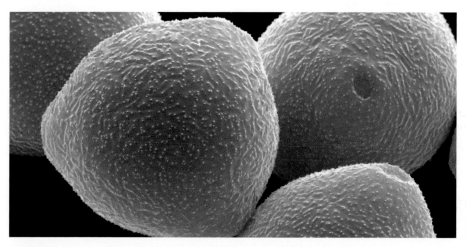

◀ **Figure 4.55** Hazel pollen grains from peat. Magnification ×2900.

Q4.23 Alder trees grow in damp soil, often near rivers, lakes and marshes. What would the appearance of a lot of alder pollen in a particular peat layer suggest about the climate at the time it was deposited?

From the abundance of pollen grains in a sample, a pollen diagram can be constructed (Figure 4.56).

Q4.24 Figure 4.56 shows a pollen diagram taken from Hockham Mere in Norfolk. Describe the changes that have occurred in the species abundance as you move forward in time towards the present.

Activity

Collect your own pollen core, without getting your feet wet, and reconstruct climate change over the past 10 000 years using virtual pollen analysis in **Activity 4.21**. **A4.21S**

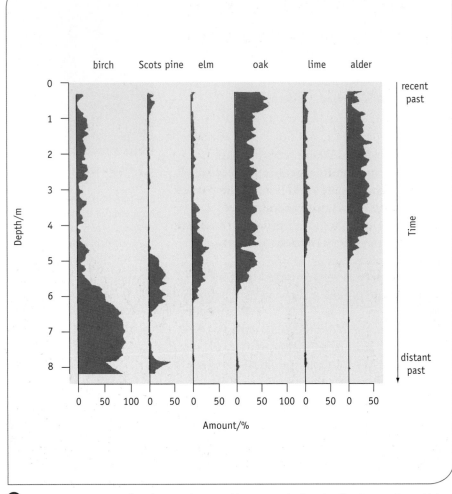

Figure 4.56 A tree pollen diagram from Hockham Mere in East Anglia. *Source: Open University.*

Bog beetles

Plants respond rather slowly to changes in climate, but insect populations respond much faster. Studies of the exoskeletons of beetles in peat can therefore give a more precise measure of climate change.

Q4.25 Different species of beetle thrive in specific temperature conditions, so their remains in peat and lake sediments can be used as indicators of past climate conditions. Table 4.2 shows data obtained in this way.
a What is the advantage of giving temperatures separately for February and July rather than as an average for the year?
b What conclusions can you draw about climate change over this period of time? Sketching a graph may help.

Table 4.2 Temperatures estimated from beetle remains preserved in layers of peat from different periods.

Period/years before present	Average July temperature/°C	Average February temperature/°C
18 000–13 000	10	−20
13 000–12 000	17	0
12 000–11 000	15	−5
11 000–10 000	10	−20
10 000–7500	16	3
7500–5000	18	5
5000–2500	17	4
2500–present	15	5

Tree-ring analysis – dendrochronology

Every year, trees produce a new layer of xylem vessels by the division of cells underneath the bark. The diameter of the new xylem cells varies according to the season when they are produced: wide vessels in spring when the tree grows quickly, followed by narrow vessels in summer. Little if any growth takes place in autumn and winter. The different widths of the vessels create a pattern of rings across the trunk which can easily be seen when a tree is cut down, with a ring for each year of tree growth (Figure 4.57). Instead of cutting down a tree to see the rings, a core sample can be taken and examined.

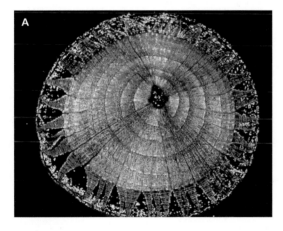
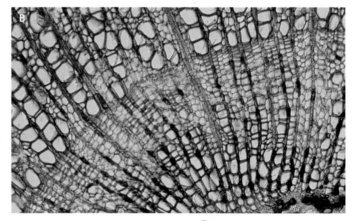

▲ **Figure 4.57 A** Cross-section of the trunk of a lime tree showing tree rings. **B** Close-up of the xylem vessels. The larger vessels are produced in spring; the narrower vessels in summer.

Q4.26 Where will the youngest rings be located, on the outside or deeper in the tree trunk?

If you cut a tree down or take a core sample in autumn 2006, the outermost ring will have come from growth in 2006. By counting inwards you can date the year each ring was formed. To date preserved trees or wood samples accurately, experts find common patterns of tree-ring growth that allow cross-dating (Figure 4.58).

If you find that the ring made in the year 1326 was wider than that made in 1327, it means that the tree grew more in 1326 than 1327; this probably indicates that the conditions in 1326 were better for tree growth. It was probably warmer or wetter (Figure 4.59). So tree rings can give not only precise dates but also strong clues about past climates. The study of tree rings is known as **dendrochronology**.

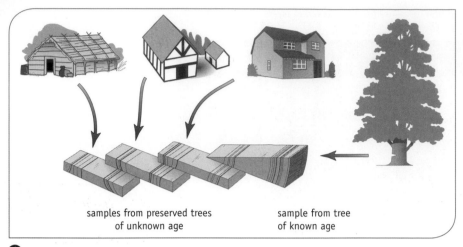

▲ **Figure 4.58** Matching the pattern of tree rings allows wood to be dated.

Q4.27 Look at the graph of tree rings in Figure 4.59. The width of tree rings is related to their conditions of growth.
a Which two separate years were particularly cold or dry?
b Which was probably the warmest or wettest?

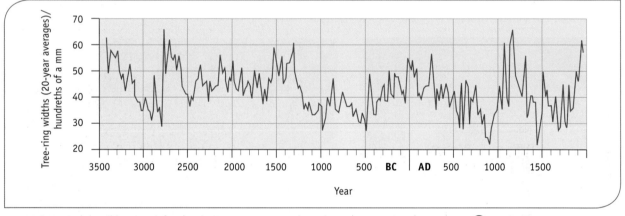

▲ **Figure 4.59** Graph showing the change in average tree-ring width with time.

Putting the data together

Historical temperature records made by people provide climate information from the present back to about 1650. Tree-ring studies extend this record back hundreds of years, 3000 years in some cases. Pollen gives us information going back some 20 000 years. Ice cores are used to find out what happened before this. As water freezes, bubbles of air become trapped within the ice. The ratio of different oxygen isotopes in the trapped air is measured, and this gives an estimate of the average air temperature when the ice was formed. The carbon dioxide concentration of the air can also be determined from these bubbles.

Combining the evidence obtained using all these techniques allows us to build up a picture of the conditions over the last 160 000 years. Figure 4.60 shows the changes in temperature over this period. It is clear that major fluctuations in climate have occurred regularly. Changes during the past century indicate that we are in a period of rapid global warming (Figure 4.61).

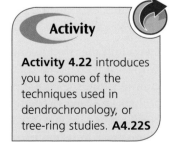

Activity

Activity 4.22 introduces you to some of the techniques used in dendrochronology, or tree-ring studies. **A4.22S**

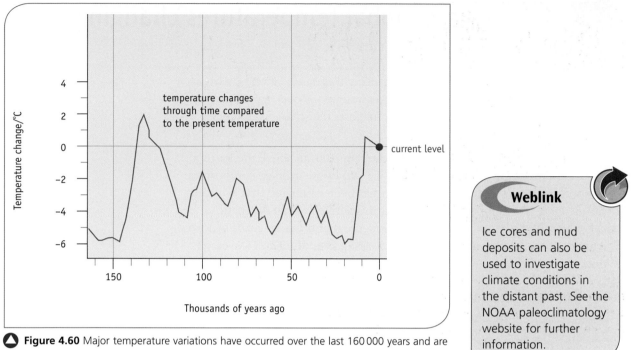

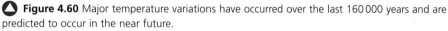

Weblink

Ice cores and mud deposits can also be used to investigate climate conditions in the distant past. See the NOAA paleoclimatology website for further information.

▲ **Figure 4.60** Major temperature variations have occurred over the last 160 000 years and are predicted to occur in the near future.

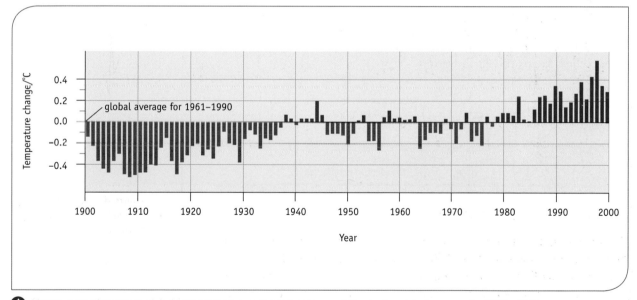

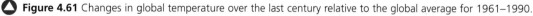

▲ **Figure 4.61** Changes in global temperature over the last century relative to the global average for 1961–1990.

4.7 Why are global temperatures changing?

Climate patterns have been fluctuating for many thousands of years, with temperature and rainfall both subject to great variations over time. There seems to be a body of reliable data which suggests that changes in the atmosphere are linked to climate change.

The atmosphere is a thin layer of gases extending 100 km above the Earth's surface and held in place by gravity. Table 4.3 shows the average composition of the bottom 15 km of the atmosphere. It has been estimated that, without an atmosphere, the temperature of the Earth's surface would fluctuate between very hot days and very chilly nights, the nights being −40 °C or lower. The atmosphere has an important role in helping keep the Earth's average temperature stable and suitable for living organisms.

Keeping the Earth warm – the greenhouse effect

The Sun radiates energy, largely as visible light, and the Earth absorbs some of this energy. The Earth warms up and in turn radiates energy back into space as infrared radiation (Figure 4.62). Some of the energy that is radiated from the Earth's surface is absorbed by gases in the atmosphere, warming it. In Figure 4.63 you can see how this is similar to the way that glass traps energy in a greenhouse. The gases in the atmosphere that stop the infrared radiation from escaping are called **greenhouse gases**. They create the **greenhouse effect** which keeps the Earth warm.

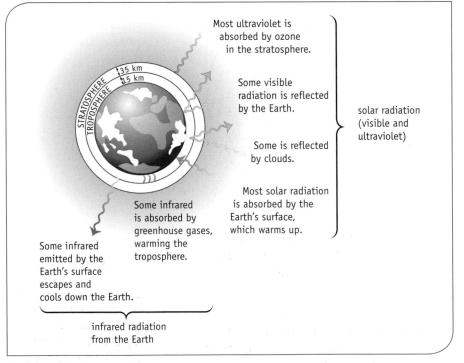

▲ **Figure 4.62** Inputs and outputs of energy to the Earth's atmosphere.

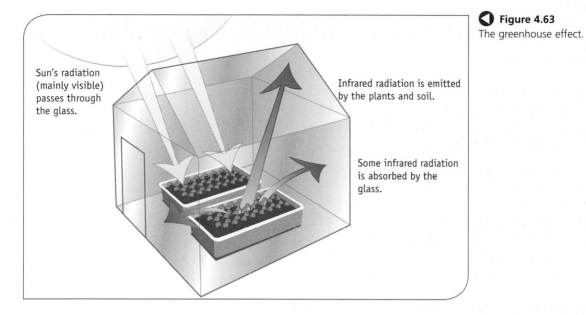

◀ **Figure 4.63**
The greenhouse effect.

Sun's radiation (mainly visible) passes through the glass.

Infrared radiation is emitted by the plants and soil.

Some infrared radiation is absorbed by the glass.

Which are the greenhouse gases?

Not all gases are greenhouse gases. In Table 4.3 you can see the relative contribution of the different gases in the atmosphere to the greenhouse effect. The 'greenhouse factor' is a measure of the greenhouse effect caused by that gas relative to the same amount of carbon dioxide, which is given a value of 1. Although carbon dioxide does not have the largest greenhouse factor it is so much more abundant than the more potent greenhouse gases that it has the largest effect, apart from water.

▼ **Table 4.3** The contribution of different gases in the bottom 15 km of the atmosphere to the greenhouse effect.

Gas	Abundance in atmosphere /% by volume	Greenhouse factor
nitrogen	78	negligible
oxygen	20	negligible
argon	1	negligible
water (gaseous)	1	0.1
carbon dioxide	3.7×10^{-2}	1
methane	1.8×10^{-4}	20
nitrous oxide	3.1×10^{-5}	310
chlorofluorocarbon CCl_3F	2.6×10^{-8}	3800

Q4.28 Look at Table 4.3. Water's contribution to the greenhouse effect can be calculated by multiplying its percentage abundance (1) by its greenhouse factor (0.1), giving a value of 0.1. Calculate the contributions to the greenhouse effect of carbon dioxide and methane.

Methane (CH_4) is produced by anaerobic decay of organic matter in waterlogged conditions, for example in bogs and rice fields. Decay of domestic waste in landfill sites and the decomposition of animal waste are also sources of methane. It is also produced in the digestive systems of animals such as cattle and released when they belch and fart. Incomplete combustion of fossils fuels also releases methane.

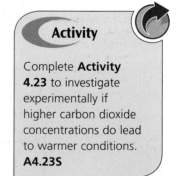

Activity

Complete **Activity 4.23** to investigate experimentally if higher carbon dioxide concentrations do lead to warmer conditions. **A4.23S**

Table 4.4 shows that since pre-industrial times the atmospheric concentration of methane has risen by 150%, although the rate of increase is declining. Methane absorbs infrared radiation more than carbon dioxide does, so it could be regarded as an even worse greenhouse gas (see Table 4.3). Unlike carbon dioxide, however, it does not stay in the atmosphere very long. In less than 12 years a methane molecule will have reacted with oxygen in the air to form carbon dioxide and water but, of course, more methane keeps being made.

Methane emissions could be reduced by better waste recycling and by using it as a biofuel (see page 210). When methane burns and produces carbon dioxide and water vapour, one greenhouse gas changes into two less serious ones. It would eventually have turned into these gases anyway, and the energy that comes from burning methane from waste can replace energy from fossil fuel.

Significant increases in carbon dioxide levels in the atmosphere through processes such as the combustion of fossil fuels have been linked to changing temperatures. The Intergovernmental Panel on Climate Change estimates that carbon dioxide in the atmosphere has risen by 31% since industrialisation in the nineteenth century (Table 4.4).

▼ **Table 4.4 A** Changes in the concentrations of key greenhouse gases. *Source: Technical summary by working group 1 of the Intergovernmental Panel on Climate Change.*

Global values	Carbon dioxide CO_2	Methane CH_4	Nitrous oxide N_2O	Chlorofluoro-carbons CFC
Pre-industrial concentration (1750)	about 280 ppm	about 700 ppb	about 270 ppb	zero
Concentration 1998	365 ppm	1745 ppb	314 ppb	268 ppt
Rate of concentration change (1990–1999)	1.5 ppm/yr[a] (0.9–2.8 ppm/yr)	7.0 ppb/yr[a] (0–13 ppb/yr)	0.8 ppb/yr	−1.4 ppt/yr
Atmospheric lifetime	5 to 200 yr	12 yr	114 yr	45 yr

[a]Rate has fluctuated between 0.9 and 2.8 ppm/yr for CO_2 and between 0 and 13 ppb/yr for CH_4 between 1990 and 1999.

▼ **Table 4.4 B** Estimated UK emissions of greenhouse gases 1990–1999/million tonnes of CO_2. *Source: Digest of Environmental Statistics March 2004, Department for Environment, Food and Rural Affairs.*

Year	1990	1991	1992	1993	1994	1995	1996	1997	1998	1999	2000	2001	2002
Carbon dioxide	602.8	606.7	591.9	576.3	572.4	563.4	583.0	558.0	560.3	554.2	557.9	571.1	551.0
Methane	3.67	3.62	3.53	3.38	3.07	3.05	2.98	2.89	2.76	2.63	2.32	2.19	2.10
Nitrous oxide	0.22	0.21	0.19	0.18	0.19	0.18	0.19	0.19	0.19	0.14	0.14	0.14	0.13

Does increased carbon dioxide cause global warming?

Compare the changes in temperature over the last 160 000 years with the changes in carbon dioxide concentration over the same periods (Figures 4.60 and 4.64) There appears to be a correlation, but is there a causal relationship? Is one (carbon dioxide) causing the other (warming)?

Comparison of the graphs does provide some support for the theory that a rise in carbon dioxide is strongly linked to the increase in global temperature; yet it does not *prove* that one causes the other. The evidence

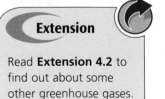

Extension

Read **Extension 4.2** to find out about some other greenhouse gases. **X4.02S**

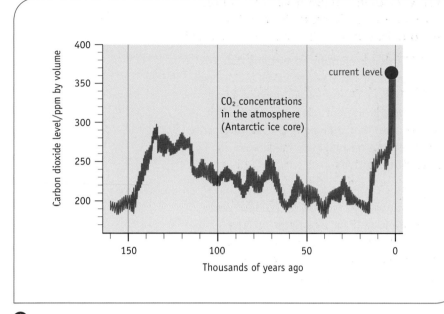

▲ **Figure 4.64** Changes in carbon dioxide concentration over the last 160 000 years. Current levels of carbon dioxide are estimated to be 30% higher than at any time over the last 400 000 years.

is still circumstantial, and it could be that some factor other than carbon dioxide concentration is also influencing global temperature. But although the evidence is circumstantial, there is so much of it that most scientists accept that a connection is likely enough for it to be taken seriously.

The rise in global temperatures due to rising levels of greenhouse gases is now a widely accepted theory. It is worth noting that scientists use the word 'theory' in a different way from how we would use it in everyday speech. 'Theory' to some people implies a speculative idea, a weak hypothesis, or an idea which lacks much evidence or reliability. In science 'theory' implies a well tested and widely accepted idea or principle supported by a great deal of evidence. In science, theories are the most reliable types of ideas we have.

A controversial issue

Global warming has been described by many people as a controversial issue. An issue is controversial when alternative points of view about it can reasonably be held. Global warming remains controversial because:

- Science cannot prove theories – scientific methods can only disprove theories. Using scientific methods an idea (hypothesis) is proposed to explain an observation, and it is then tested. If the results disprove the idea it is rejected; if the results support the idea it does not actually prove it – there could be alternative explanations.
- There is incomplete knowledge of how the climate systems of our planet work and the datasets used in making predictions about climate change have their limitations. For example, there is no way to measure precisely how much carbon dioxide is added to the atmosphere by fossil fuel combustion. This doesn't mean that the numbers reported are mere

Activity

In **Activity 4.24** you compare carbon dioxide levels and global temperatures. **A4.24S**

Weblink

Visit the British Antarctic Survey website to find out about ice core records for carbon dioxide over the last 740 000 years.

guesses. They are estimates based on scientifically defensible procedures. But some people use this uncertainty to dismiss the link between rising carbon dioxide levels and global warming.

Despite much controversy, there is widespread scientific consensus that temperatures are rising and that rising levels of greenhouse gases are at least partly responsible. This consensus results from a gradual build-up of a large body of scientific evidence supporting the theory. But some people do not agree with the consensus, suggesting instead that the changes in temperature that are currently being observed are part of a natural cycle of climate variations or are due to changes in the Sun's activity.

When presenting and interpreting scientific evidence almost everyone is influenced by his or her own particular values and viewpoint. Political and economic considerations may affect how individuals, organisations and countries interpret the evidence. Where cuts in carbon dioxide emissions to reduce global warming might harm business, a company or country may not accept the causal link between the two factors. In the US the oil and power industries support organisations that advocate such scepticism. Some of these organisations go further and suggest that rising carbon dioxide levels will be a benefit through increased plant productivity.

Ethical arguments are often quoted when considering the issue of global warming. These arguments include:
- We all have the right to choose for ourselves whether we use fossil fuels to achieve a good standard of living.
- We have a duty to allow others to improve their standard of living (which is often equated with industrialisation).
- We have a duty to preserve the environment for the next generation.

Q4.29 Suggest a further ethical argument based on the framework of utilitarianism.

The fact that climate change cannot be conclusively proved leaves us with a dilemma. If we therefore do nothing, the consequences could be very serious. Scientists use mathematical models to try and predict what these consequences might be.

> **Activities**
>
> **Activity 4.25** allows you to critically evaluate views expressed in articles about global warming. **A4.25S**
>
> In **Activity 4.26** you debate the different views expressed on global warming. **A4.26S**

Did you know? Kyoto Protocol

The Kyoto Protocol was an international agreement to cut back on greenhouse gas emissions across the globe. It came about as a result of discussions at the Earth Summit in Rio, Brazil in 1992 and was agreed as a formal strategy in Kyoto, Japan in 1997 (Figure 4.65). It has been very difficult to get countries to agree on just how much the emissions need to be cut in order to prevent serious damage to our climate. However, in 2001 agreements were reached and 178 countries signed the treaty.

There were two major exceptions: the USA and Russia, who failed to sign up. This was partly because American politicians accepted the arguments of big businesses, particularly the oil industry, that the Kyoto Protocol would be bad for the American economy. As the world's largest producers of carbon dioxide, this does cause some concern. For the Kyoto treaty to be effective we need to include the USA in the long term.

In November 2004 Russia signed the treaty which brings it into force without the USA. Industralised countries that signed the treaty have agreed to cut their combined emissions to 5% below 1990 levels by 2008–2012.

Weblink

Find out who agreed to the Kyoto Protocol by visiting the UN framework convention website. Find out about international developments since Kyoto by doing some web research including a visit to the World Resource Institute's website.

▲ **Figure 4.65** The Kyoto Protocol was agreed at the UN Climate Conference held in Kyoto, Japan in 1997.

4.8 Predicting future climates

Enhanced global warming due to rising carbon dioxide levels is only one of several factors likely to affect the future climate. Scientists use computer models to study the interaction of many factors in an attempt to make predictions about likely climate change.

Making mathematical models

A really reliable model

If a square has one side of 10 cm can we 'predict' the length of the other three sides? That one's easy. By definition all the sides of a square are equal in length, and so the sides are each 10 cm long. This model gets the right answer every time!

Extrapolation

Figure 4.66 shows the actual data for atmospheric carbon dioxide covering the period 1958 to 1988, recorded in the field. Since we are interested in the long-term view we can ignore the annual fluctuations (due to different levels of photosynthesis in winter and summer), and apply a smoothed best-fit line. This produces a steep upward trend in Figure 4.66.

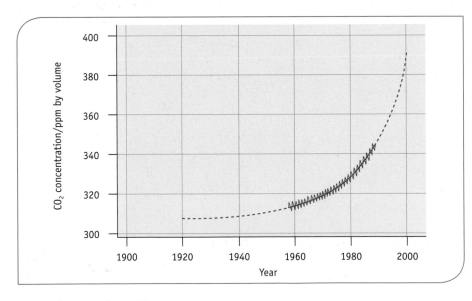

Figure 4.66 Changes in atmospheric carbon dioxide 1958–1988 measured at Mauna Loa observatory in Hawaii, with extrapolation back to 1920 and forward to 2000.

Q4.30 Use Figure 4.66 to estimate the carbon dioxide concentration in:
a 1940 **b** 1995.

In Figure 4.66, the smoothed best-fit line is extended (dotted line) back to 1920 and forwards to 2000. Extending a line is called extrapolation. In extending the lines we make the assumption that:
- we have enough data to establish the trend accurately
- present trends continue.

The easiest line to extrapolate is a straight one – you just use a ruler. To extrapolate a curve you need to identify the shape of the curve. Computers can extrapolate curves mathematically more exactly than humans can.

An extrapolation is often the basis for predictions; the graph in Figure 4.66 acts as a mathematical model of carbon dioxide change. However, 2000 has come and gone without the mean carbon dioxide level reaching the level predicted by Figure 4.66. We can therefore conclude that predictions based on this model are not very accurate. This Mauna Loa data is the longest set of direct measurements of atmospheric carbon dioxide concentrations in existence, but estimations based on this data tend to turn out too high (see 'The mystery of the disappearing carbon dioxide', page 209).

The present trend for increasing carbon dioxide may not continue. For example, if steps were taken to reduce CO_2 emissions, the efficiency of petrol and diesel engines were improved, and the rising cost of energy encouraged people to improve the insulation of their homes, then 'present trends' might not continue. On the other hand, increases in living standards in countries such as China and India (which between them have over 2 billion people) might cause atmospheric CO_2 levels to rise even faster than present trends suggest.

Taking into account many factors at once

If we could understand all the key factors that affect climate and how they interact, we could produce a computer model which is much more mathematically sophisticated than the graph in Figure 4.66, giving us the opportunity to make more reliable predictions.

Modelling climate change is a very complicated business, and carbon dioxide concentration is only a part of the story. Many factors are involved, and if one or more is missed out of a model then its predictive accuracy will be reduced. It is also not enough to include all the factors; the model must take into account interactions between them.

Carbon dioxide is an important factor because it is one which is changing at present and is one which humans can influence. But it isn't by any means the only one. Other factors that may affect climate change include:
- other greenhouse gases such as methane, CFCs and nitrous oxide (N_2O)
- aerosols – extremely small particles or liquid droplets found in the atmosphere
- the degree of reflection from those parts of the Earth's surface which are free of ice and snow
- the fraction of the Earth covered with ice and snow
- the extent of cloud cover.

The Hadley Centre for Climate Prediction and Research, part of the UK Met Office, has several climate models which they use to predict the surface temperatures, precipitation, soil moisture content, sea-level change, sea-ice area, and sea-ice volume at the end of the twenty-first century. The maps in Figure 4.67 assume that no measures are taken to reduce greenhouse gas emissions and that there is mid-range economic growth resulting in a doubling of emissions of greenhouse gases over the course of the twenty-first century.

Activity

Activity 4.27 provides the opportunity to make use of a computer model of climate change. **A4.27S**

Weblink

You can view more information about the Hadley Centre and their climate models by visiting their website via the Met Office site. This also provides lots of links to other useful sites.

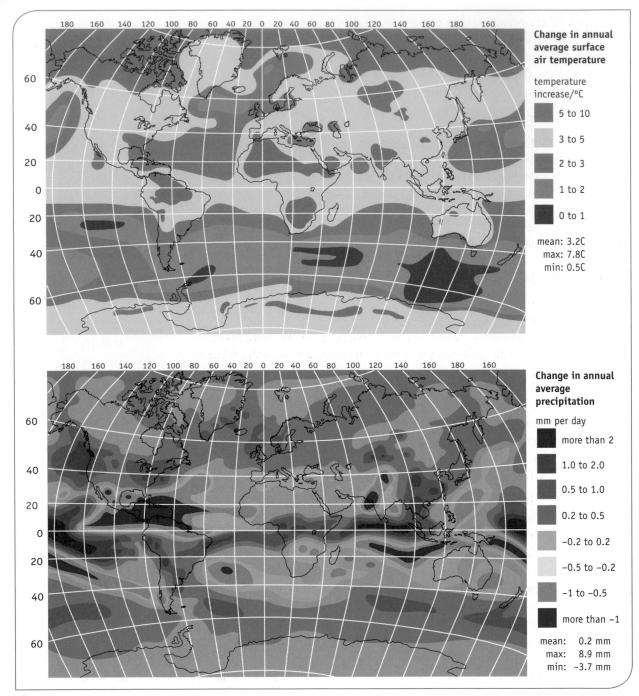

🔺 **Figure 4.67** Maps based on the Hadley Centre computer model predictions for climate at the end of the twenty-first century relative to current climate, defined as 1960–1990. *Source: Met Office.*

Don't expect models to be perfect

Several major climate models are currently in use around the world. They do not always give the same answers to the same question, and sometimes all the models 'get it wrong'. The predictions may be incorrect because of:

• limited data
• limited knowledge of how the climate system works

- limitations in computing resources
- failure to include all factors affecting the climate
- changing trends in factors included, e.g. faster than expected loss of snow and ice cover or greater carbon dioxide emissions.

The models are, however, continually improving, using bigger datasets and incorporating more factors and more sophisticated interactions.

Taking data from years past, feeding the data into the models and seeing whether the predictions match what actually happened can test climate models. Of course, even if a model works when tested with old data, we can't be sure that it will be reliable in future. For one thing, other factors may change in ways we never expected. Models are not expected to predict the future precisely, but to make the best prediction based on all the evidence available.

Climate model predicts a colder UK

Winters in Britain are much milder than in other places on a similar latitude. For example, Goose Bay in Canada is the same latitude as Manchester UK, but the winter temperature in Goose Bay can drop to $-50\,°C$. Apart from Alaska, the whole of the US lies south of Britain. The UK has milder winters than much of the US, because of the Gulf Stream and its extension the North Atlantic Drift which brings warm water from the Gulf of Mexico to north-west Europe. In the North Atlantic, surface water cools, contracts and sinks. This cooled water flows deep below the ocean surface and eventually goes back to the Gulf of Mexico where it is warmed again, expands and rises to the surface. This sinking of water in the North Atlantic acts as a 'pump', causing warm surface water to flow northwards.

Global warming has caused a lot of ice in Arctic areas to melt, and the extent of the ice sheet is reducing. The melting ice means an influx of fresh water into the North Atlantic. Near freezing point, fresh water does not behave like salt water. At $0\,°C$, fresh water starts to freeze and expands. This water rises to the surface rather than sinking as salt water of the same temperature does. At some critical point, the North Atlantic Drift could break down and no longer bring warm water from the Gulf of Mexico. If this were to happen, the average temperature across Britain would fall. This has not happened yet and may never do so, but there is evidence that the salt concentration of the North Atlantic has been falling over the past two decades. We can't be sure it's due to global warming, but this is certainly a possible cause which needs to be taken very seriously.

Stefan Rahmstorf of the Potsdam Institute for Climate Impact Research, using a computer model, has predicted that global warming could result in a reduction of surface temperature in north-west Europe. A $5\,°C$ fall in the mean surface temperature is predicted, which is quite substantial – not quite an Ice Age but getting that way.

So for Britain there are two scenarios which we could ask our climate models about. What would happen if surface temperature fell by $5\,°C$, and what would happen if it rose by $5\,°C$? Each could form the basis for a disaster movie.

4.9 The carbon dioxide balance

Key biological principle: The carbon cycle

The amount of carbon dioxide in the atmosphere, approximately 0.03–0.04%, is maintained by a balance between the processes that remove carbon dioxide from the air and those that add carbon dioxide to it. This circulation of carbon is known as the **carbon cycle** (Figure 4.68).

▼ **Figure 4.68**
The carbon cycle.

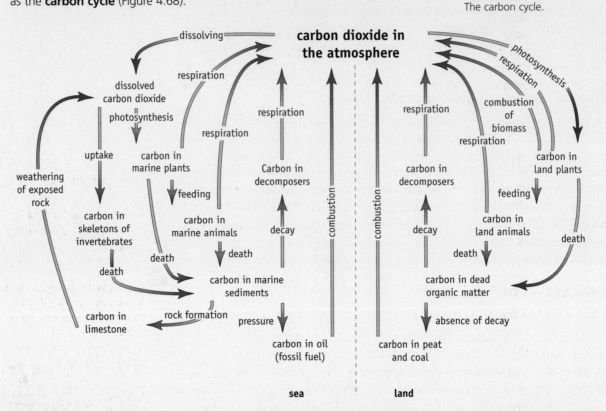

Q4.31 For each definition below, decide which term in the carbon cycle shown in Figure 4.68 is being described.

a The reduction of carbon dioxide to organic substances such as sugars, using light energy, carried out in the chloroplasts of plants and in the cells of some microbes. Some of these sugars are used by the plant directly to provide energy through respiration, some are converted to starch for storage, and the rest are used to make proteins, nucleic acids, lignin and cellulose. Oxygen is given off as a waste product.

b The total amount of living material per unit area. It includes animal and microbial (bacterial and fungal) mass but by far the biggest proportion of it in an ecosystem is plant material. This includes the wood and foliage of trees and grass, and the countless millions of microscopic phytoplankton (plant plankton) which account for most of the photosynthesis of the oceans.

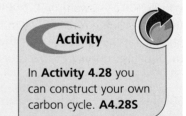

Activity

In **Activity 4.28** you can construct your own carbon cycle. **A4.28S**

c The oxidation of organic substances, such as sugars, to simpler inorganic compounds, such as carbon dioxide and water, with the release of biologically available energy. Carried out by plants and animals all the time.
d Sooner or later plants and animals die and, along with animal faeces, are broken down. Carbon dioxide is released during the breakdown by microbes (types of fungi and bacteria) when they use the dead organic matter as a food to obtain energy by respiration.
e Oxidation of organic molecules outside living cells with the formation of carbon dioxide and water.

The role of microbes in the carbon cycle is vital. If there were no microbes to recycle plant biomass, by giving out carbon dioxide during respiration, most of the carbon dioxide in the air would be used up in photosynthesis in less than 10 years. This would mean the greenhouse effect would be greatly reduced, and the environment would soon become too cold to support life (not enough global warming). Photosynthesis would greatly slow down as temperatures fell, and so there would be slower plant growth and less food for animals.

Out of balance

The fact that carbon dioxide concentrations have been rising for over 100 years indicates that the carbon cycle is not in balance. Two factors likely to be responsible for this are:
- combustion of fossil fuels
- deforestation.

Combustion of fossil fuels

We are currently burning up fossil fuels much faster than they are forming. Table 4.5 shows the large amounts used in the UK and this represents a significant addition to the carbon dioxide already in the atmosphere.

The fossil fuel coal was formed from the wood of trees which lived millions of years ago, and so is a product of photosynthesis. The wood did not decay, releasing the carbon dioxide, because when the trees died they fell into conditions where fungi and bacteria could not grow, such as the anaerobic conditions of a swamp (Figure 4.69). Carbon remains locked in this coal instead of being returned to the air. The coal thus represents a **carbon sink**, and its accumulation involves the net removal of carbon dioxide from the air over millions of years.

Figure 4.69 A swamp amongst prehistoric forest where coal might have formed 250 000 000 years ago.

Extension

Read **Extension 4.3** for more detail about decay.
X4.03S

▼ **Table 4.5** Energy use in the UK between 1990 and 2003. *Source: Department of Trade and Industry.*

Year	Million tonnes of oil equivalent (temperature corrected)*
1990	221.6
1991	221.4
1992	220.6
1993	222.5
1994	221.5
1995	226.6
1996	230.1
1997	232.6
1998	235.8
1999	235.7
2000	237.9
2001	238.7
2002	236.2
2003	236.8

*Energy consumption for particularly cold and mild winters is adjusted to give values that can be compared directly.

When we extract and burn fossil fuel, the carbon released as carbon dioxide has been out of circulation for millions of years. Since the late eighteenth century in Britain we have been burning more and more fossil fuel, setting free carbon which has taken millions of years to accumulate.

Deforestation

Most mature forests are very stable ecosystems. They do not accumulate additional biomass or become fossil fuel, so they are not net absorbers of CO_2. Carbon dioxide uptake by photosynthesis is expected to be equal to the release of CO_2 due to respiration (including decay). If the forest were cut down, photosynthesis would drop; and although in the long term respiration would also drop, in the short term more carbon dioxide will be released than absorbed.

Often only the large pieces of timber from big trees of the species being harvested are removed. The rest includes large amounts of discarded branches, small trees, shrubs and other plants which are of no use to the loggers. These are either left to rot away or burnt (Figure 4.70), releasing carbon dioxide.

▲ **Figure 4.70** A recently felled rainforest. If the discarded wood were not burnt it would still decay and release just as much carbon dioxide.

What else could upset the carbon dioxide balance?

There are other factors affecting carbon dioxide levels in the air, but these are not thought (at least at present) to be significantly upsetting the equilibrium of the carbon cycle:

- Volcanoes may release CO_2 – an increase in volcanic activity in future could make a bigger difference.
- Carbon dioxide is continually being lost to sediments in the ocean by various processes such as the incorporation of carbon into the calcium

carbonate skeletons and shells of marine organisms, but this is balanced by various erosion processes.
- An increase in acid rain might increase the rate at which CO_2 is released by erosion of limestone, but this is not thought to be a major factor upsetting the balance at present.

The mystery of the disappearing carbon dioxide

It is estimated that burning fossil fuels adds 5.4×10^{12} kg of carbon per year, and that deforestation adds approximately 1.6×10^{12} kg of carbon per year. However, the actual increase in carbon dioxide is only 3.0×10^{12} kg of carbon per year. Some carbon dioxide seems to be disappearing. There is still quite a lot of debate about why the carbon dioxide level doesn't rise as fast as we would expect. One explanation, suggested by the workings of the carbon cycle, is simply that more carbon dioxide means more photosynthesis.

In order to investigate the movement of carbon through an ecosystem, carefully controlled plots can be set up in artificial chambers called ecotrons. Results from these suggest that when CO_2 levels are raised, more carbon is stored in other components of the carbon cycle, for example within the soil as dissolved organic components. There is also evidence of a rise in the rate at which carbon dioxide is dissolving in the ocean. One reason for this could be increased photosynthesis by massive blooms of algae.

Maintaining the balance

We can help maintain the carbon dioxide balance by:
- using biofuels
- reafforestation.

Using biofuels

A **biofuel** is any source of energy produced, directly in plants or indirectly in animals, by recent photosynthesis. This provides a *renewable* energy source and is *carbon dioxide neutral*. For example, when wood burns, the carbon dioxide released to the atmosphere merely replaces the carbon dioxide that was previously absorbed through photosynthesis and which would have eventually been released through the processes of decay in any case. The release of carbon dioxide in combustion does not therefore cause a net increase. The exception comes, of course, when carbon dioxide is released in the process of transporting the biofuel from where it is made to where it is used.

Q4.32 Why is the use of biofuel carbon dioxide neutral?

Examples of biofuels include wood, straw, dried chicken litter, vegetable oil and methane. In 2004, Drax, Europe's largest coal-fired power station, started testing the use of fast-growing willow biomass mixed with coal. Another example of biofuel is ethanol from the fermentation of any kind of cheap and locally available sugar.

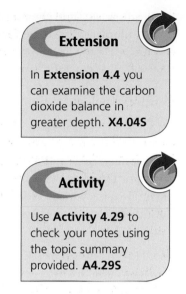

Extension

In **Extension 4.4** you can examine the carbon dioxide balance in greater depth. **X4.04S**

Activity

Use **Activity 4.29** to check your notes using the topic summary provided. **A4.29S**

Weblink

You can find out more about renewable energy by taking a virtual tour of the Centre for Alternative Technology on their website.

In Brazil, waste from the refining of sugar cane is used, and the resulting alcohol added to petrol to make gasohol. In the UK we could grow sugar beet to provide the sugar to make alcohol as biofuel. Vegetable oil from, for example, sunflower seeds (Figure 4.71) is a very energy-rich biofuel which can be used as a substitute for diesel in motor vehicles.

At present the use of plant-derived alcohol and oil to power vehicles is limited because it remains cheaper to use petrol or diesel. However, rising oil prices and improved technology may make biofuels more cost effective in the future. If production costs can be kept low this will help make biofuels a sustainable energy source.

Methane produced from anaerobic fermentation of human sewage can be used to generate enough electricity to make a modern sewage treatment plant energy self-sufficient. Methane can also be produced from domestic waste and from animal slurry. Methane produced in this way may be referred to as biogas and, like solid biofuel, when burnt is CO_2 neutral, unlike fossil fuel methane from the North Sea.

Reafforestation

In a newly planted forest all the trees are young; all are growing rapidly, turning carbon dioxide into wood. There is very little old wood and relatively little decay and so respiration will be less than photosynthesis, meaning that the system is a net absorber of carbon dioxide. As the plantation gets older the system will move towards a balance between photosynthesis and respiration and will no longer be a net absorber. It becomes a carbon sink with carbon locked up in the biomass.

The Earth's forests could soak up extra carbon dioxide as increased CO_2 concentrations and higher temperatures stimulate photosynthesis, leading to extra growth. However, there will be a limit to how much or for how long the world's forests can soak up extra carbon dioxide. In the long term more vegetation means more food, therefore more animals and more decay-causing microbes, hence more respiration producing more carbon dioxide. There is an upper limit to tree growth, and the land available for forests is also limited. It has been suggested that an increase in mean global surface temperatures above a certain level (perhaps an increase of 5 or 6 °C) could reduce water availability in rainforests. This would cause them to go into decline and accelerate further increases in atmospheric carbon dioxide, resulting in even further global warming.

Both the microscopic algae in the oceans and the Earth's forests could act as carbon sinks, soaking up extra carbon dioxide and incorporating it into new biomass. Planting trees, along with a range of other actions such as less use of fossil fuels, may help slow down further increases in atmospheric carbon dioxide concentrations.

Congratulations – you have completed the AS!

△ **Figure 4.71** Sunflower plants can be grown to supply oil as fuel.

Topic test

Now that you have finished Topic 4, complete the end-of-topic test.

Answers to in-text questions

Topic 1

Q1.1 Movement of oxygen; carbon dioxide; and other products carried by blood; relies on diffusion in animals with an open circulatory system; diffusion is only fast enough for small organisms;

Q1.2 Blood can pass slowly though the region where gaseous exchange takes place; maximising the transfer of oxygen and carbon dioxide; and then be pumped vigorously round the rest of the body; enabling the organism to be very active;

Q1.3

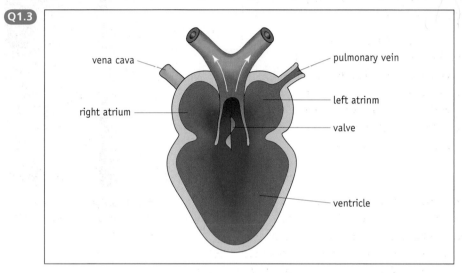

Q1.4 Some mixing of oxygenated and deoxygenated blood; in the ventricle;

Q1.5 Thick layer of mainly elastic fibres; and smooth muscle; to allow expansion of diameter of artery; surrounded by thick layer of mainly collagen fibres; to aid in contraction of diameter of artery;

Q1.6 The blood pressure in the left ventricle falls below that in the aorta; leading to the closure of the semilunar valve between the left ventricle and the aorta; the blood pressure in the right ventricle falls below that in the pulmonary artery; leading to the closure of the semilunar valve between the right ventricle and the pulmonary artery;

Q1.7 There are five squares between the QRS complexes, thus each heartbeat lasts one second; which gives a heart rate of 60 beats per minute; the heart rate can be calculated by dividing 300 by 5 = 60 beats per minute;

Q1.8 **a** No clear pattern of P wave, QRS complexes and T waves; irregular timing and strength of waves; **b** Very high heart rate; abnormal T wave;

Q1.9 Between the sinoatrial node and the atrioventricular node; or at the Purkyne fibres;

Q1.10 Congenital heart problem; extreme exhaustion; extreme dehydration; use of certain illegal drugs;

Q1.11

Cause	Probability	
	Percentage	**Per 1 000 000 people**
heart disease	0.23%	2 300
lung cancer	0.06%	600
road accidents	0.006%	60
accidental poisoning	0.0017%	17
injury purposely inflicted by another person	0.000 49%	4.9
railway accidents	0.000 07%	0.7
lightning	nearly 0	0.1

Q1.12 Ten students out of 1300 ($\frac{10}{1300}$) is less than 1%; assumptions: the same exposure for each student; e.g. the same frequency of swimming for each student, the same length of each term with the same frequency of swimming each term and the same viral load in and around the pool each term;

Q1.13 **a** No causal link; hot weather increases the number of ice cream sales and the number of people going swimming off shark-infested beaches;
b No causal link; both these variables increase with age;
c Causal link; smoking greatly increases the chance of developing lung cancer because of tars and other substances in the smoke;

Q1.14 It increases; greatly;

Q1.15 Not necessarily; it might be that behaviours early in life greatly affect one's subsequent chances of dying from cardiovascular diseases;

Q1.16 No; between the ages of 10 and 79 males are more likely to die from cardiovascular disease than are females; after this, the greater number of deaths among women is simply because they greatly outnumber men;

Q1.17 The data provide some support for the view that until the menopause a woman's reproductive hormones offer her protection from coronary heart disease in that deaths from cardiovascular diseases increase more steeply among women over the age of 50 than they do among men; appropriate reference to data; but there are other possible explanations for this; so it would be premature to draw such a conclusion with any confidence;

Q1.18 Blood not pumped at normal rate to lungs; shortness of breath;

Q1.19 **a** Approximately 480 000 calories; **b** Approximately 480 Calories;

Q1.20 Hydrogen and oxygen are always found in a 2:1 ratio/in the same ratio as in water; the ratio of carbon to H_2O varies;

Q1.21 30.1, moderately obese;

Q1.22 The risk of death from coronary heart disease steadily rises; from around 6.5 deaths per 1000 men a year with a serum cholesterol level of between 4.1 and 4.9 mmol/l; to around 11.5 deaths per 1000 men a year with a serum cholesterol level of between 8.1 and 9.6 mmol/l;

Q1.23 The fact that pre-menopausal women generally have higher HDL:LDL ratios than men would be expected to lead to their having lower rates of coronary heart disease;

Q1.24 LDL levels in the blood will fall; less saturated fat triglycerides in the blood to combine with cholesterol and protein to form LDLs;

Q1.25 Antioxidants protect against radical damage; radicals are highly reactive; and can damage many cell components; they have been implicated in the development of heart disease and some other diseases;

Q1.26 People underestimate the risk associated with high cholesterol; they are unwilling and find it difficult to make the lifestyle changes needed to lower cholesterol;

Q1.27 People who have had a heart attack may be more motivated to follow a strict diet; they probably started with higher blood cholesterol levels making it easier for them to reduce these; they may live in institutions where control over diet is greater;

Q1.28 Cholesterol levels in the blood at the time of the stroke, blood pressure at the time of the stroke, report of the state of the artery when surgeons operated, genetic screening for the gene;

Q1.29 Eat a healthy diet; a diet low in saturated fats, low in salt, high in antioxidants; take regular exercise; do not smoke; avoid stress;

Topic 2

Q2.1 Pathogenic microorganisms have time to multiply, resulting in sickness;

Q2.2 Acid in the stomach kills the microorganisms;

Q2.3 For A, SA = 6, V = 1, SA:V = 6; for B, SA = 24, V = 8, SA:V = 3; for C, SA = 96, V = 64, SA:V = 1.5;

Q2.4 **a** Its surface area increases by a factor of 4; **b** Its volume increases by a factor of 8; **c** Its surface area to volume ratio halves;

Q2.5 The surface area to volume ratio would continue to fall; the organism would not be able to exchange enough substances to survive;

Q2.6 Hippopotamus;

Q2.7 For D, SA = 34, V = 8, SA:V – 4.25; for E, SA = 28, V = 8, SA:V – 3.5;

Q2.8 Volumes are all the same, 8; but the more elongated the block the greater the surface area and thus the larger the surface area to volume ratio;

Q2.9 Tapeworm;

Q2.10 Desiccation/dehydration problems; surface also has protective function;

Q2.11 I;

Q2.12 (Two of) lungs; gut; kidney; capillaries;

Q2.13 They are carried in the bloodstream;

Q2.14 The kinks in the fatty acids prevent them lying very close together; this creates more space in which the molecules can move;

Q2.15 **a** Diffusion; **b** Active transport; **c** Active transport; **d** Facilitated diffusion/channel protein; **e** Active transport; **f** Osmosis;

Q2.16 Salt is normally reabsorbed from sweat using the CFTR channel; with CF this does not function; so the salt is not absorbed making salty sweat;

Q2.17 It is a polymer; of nucleotides;

Q2.18 A and G both have a two ring structure whereas C and T have only one ring; the bases pair so that there are effectively three rings at each of the rungs of the DNA molecule; making the molecule a uniform width along its whole length;

Q2.19 T A G G G A C T C C A G T C A;

Q2.20 **a** Messenger RNA; **b** Transfer RNA;

Q2.21 **a** TCA; **b** AGU; **c** UCA; **d** Serine;

Q2.22 TGG;

Q2.23 A U G U A C C U A A G G C U A;

Q2.24 5;

Q2.25 UAC; AUG; GAU; UCC; GAU;

Q2.26 G T C A G T C C G;

Q2.27 **a** The sixth base is G rather than T; **b** The second base is A instead of T; **c** The fifth and sixth bases are inverted; **d** Omission of the fourth base, A; **e** An additional base is added after the fifth base;

Q2.28 The 507th triplet now reads ATT; this codes for the same amino acid isoleucine;

Q2.29 In situations 1 and 2 both parents are probably carriers and there is a 1 in 4 chance that any child they have has CF; if the father in situation 3 is not a carrier then none of the woman's children will get the disease; they will all be carriers receiving a defective allele from her and a 'normal' allele from their father; if these children were to have children with a carrier there would be a 1 in 4 chance that each child would have CF;

Q2.30 The test will not be completely reliable (it is only about 80–85% sensitive); there should not be false positives but there will be false negatives where an individual has a CF mutation but does not have one of the mutations which the test can detect;

Q2.31 Yes; if the parents believe for religious or other reasons that abortion is wrong; if they consider that the risk of miscarriage is too high; if they want to have the baby even if it will have cystic fibrosis;

Q2.32 Have a child accepting the 1 in 4 risk of its inheriting CF; use artificial insemination by donor to avoid the risk; use *in vitro* fertilisation and have the embryo screened before implantation; use prenatal screening and decide whether to continue with the pregnancy if the fetus has the disease; decide not to have children;

Topic 3

Q3.1 Cell surface membrane; smooth ER; smooth ER; mitochondrion; smooth ER, rough ER; nuclear membrane; nucleolus; nuclear membrane; rough ER; Golgi apparatus; cell surface membrane;

Q3.2 Lysosome; free ribosomes;

Q3.3 **A** nucleolus; **B** nuclear pore; **C** nuclear membrane; **D** mitochondrion; **E** smooth endoplasmic reticulum;

Q3.4 16 (= 2^4); 32 (= 2^5);

Q3.5 G1 = 8 hours; S = 4 hours; G2 = 8 hours; division = 4;

Q3.6 Ribosomes; endoplasmic reticulum;

Q3.7 Condensing allows the DNA molecules to move around the cell without getting tangled up;

Q3.8 The pores in the nuclear envelope are not large enough for whole molecules of DNA to pass through; once it disintegrates the chromosomes can move freely through the cell;

Q3.9 Mitosis is a continuous process but a slide or photograph is a static image; the shorter the stage, the less likely a cell is to be in that stage when the action is 'stopped' as the slide is made or the photograph taken; it can be deduced that the fewer cells seen in any one stage, the quicker that stage is; anaphase is shorter than telophase because there is only one cell in anaphase and three in telophase; a larger sample would be needed to give a more accurate timing;

Q3.10 **a** Asexual; sexual; **b** Sexual; asexual; **c** Sexual; asexual; **d** Sexual; asexual; **e** Asexual; sexual; **f** Asexual; sexual; **g** Asexual; sexual;

Q3.11 G1; G2;

Q3.12 The adult cell providing the genetic information to create Dolly was a specialised mammary gland cell; the successful birth of Dolly suggests that the cell must have contained all the information for making a complete organism;

Q3.13 If created in the active form it would initiate changes within the cell where it was made;

Q3.14 **a** Growth factor; acts as a signal protein; activates protein transcription factors; which switch on genes for proteins involved in bone formation; **b** Acts as a repressor molecule; may attach to promoter region for the BMP-4 gene or attach to transcription factors needed for transcription of the BMP-4 gene;

Q3.15 Protein A activates transcription factors that switch on genes coding for proteins that make up the structures at the front of the fly; the genes are only switched on where there is a high concentration of protein A;

Q3.16 **a** The proteins act as transcription factors; they bind to the DNA; and allow transcription of the gene that codes for the structures in a particular segment; **b** Mutation in the homeobox gene; so incorrect protein produced; the protein produced is the transcription factor that switches on the leg genes rather than the antenna genes;

Q3.17 Reasons 1 and 2 are the result of genotype; 3 to 6 are environmental;

Q3.18 Tyrosine is turned into melanin by a series of (bio)chemical reactions, each catalysed by a different enzyme; if any of the enzymes is missing or damaged, melanin cannot be made; mutations can occur in the genes for any of these enzymes, so there are several different mutations causing albinism;

Q3.19 The tips are slightly cooler and so the enzyme remains active; melanin is made and tips are darker;

Q3.20 Humans with this condition have heat-sensitive tyrosinase (like Himalayan rabbits and Siamese cats); so in the warmer parts of the body, such as the armpits, tyrosinase does not function and the hair is white; on the cooler surfaces of the body, tyrosinase works and the hair is dark;

Q3.21 MSH activates melanocytes by attaching to MSH receptors on the cell surface; without these receptors, MSH has no effect and no melanin is made in the hair follicles;

Q3.22 In humans sunlight stimulates MSH receptor production; there is more sunlight in the summer;

Q3.23 Each time DNA replicates before mitosis, a few errors will be made (mutations); the higher the rate of mitosis the more likely that mutations will occur in genes controlling cell activity, which can result in cancer;

Q3.24 When the embryo grows into an adult, by mitosis, the cells giving rise to the ovaries or testes may have the DNA error; thus gametes with faulty DNA could form; the cancer-causing error could be passed on in these gametes to the next generation;

Q3.25 If the single normal allele becomes damaged, a person with only one normal allele would no longer produce p53; whereas if someone with two normal alleles gets one of them damaged, they would still make p53 and be protected from cancer;

Q3.26 If chemotherapy and radiotherapy work by activating the genes for p53; they will have no effect if the genes controlling p53 synthesis are not working;

Q3.27 It is not their fault that they have the gene; most of us would prefer a society where such economic risks were spread across all of us; such discrimination would reduce the number of people going for testing; which could be fatal for them; having the gene does not mean you will necessarily develop the disease;

Q3.28 Eugenics means deliberately altering the genetic composition of the next generation; this happens if embryos containing particular genes are aborted or implanted in IVF; and also when a gamete donor is selected on the basis of sperm viability, intelligence, race or other characteristic;

Topic 4

Q4.1 **a** Nucleus; **b** Chloroplasts or mitochondria; **c** Golgi apparatus; **d** Cell surface membrane; **e** Vacuole;

Q4.2 One of the molecules will have to be inverted; so that the two –OH groups lie alongside each other and can react;

Q4.3 It links the –OH group on the first carbon of one glucose with the –OH attached to the fourth carbon of another glucose;

Q4.4

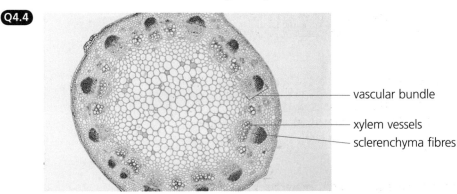

— vascular bundle

— xylem vessels
— sclerenchyma fibres

Q4.5 Release of enzymes from within the vacuole may digest cell contents;

Q4.6 Active transport;

Q4.7 Strength suited to use; length; ease and reasonable cost of extraction; durability; resistance to decay;

Q4.8 These parts come into greater contact with the soil; so are at greatest risk of microbial infection;

Q4.9 He tested the drug on patients with symptoms of the disease; recorded any side effects; used standard procedure to discover the effective dosage; slowly increased the dose until patients experienced diarrhoea and vomiting and then reduced the dose slightly; recorded all results meticulously and published results;

Q4.10 **a** It is difficult for those involved in the production of the new compound to be completely objective about it;
b Phase III is more rigorous; larger sample size; double-blind; use of statistics;
c Yes; because it is not known for certain if the new compound is effective; patients freely consent to participate in trials;
d The influence of the mind over the body;

Q4.11

1 Seeds produced in summer or autumn will be chilled over winter ready to germinate the following spring when it is warmer;
2 Seeds germinating after a fire will experience less competition;
3 Breaking down the seed coat physically or chemically may take several months and delay germination until the following spring;
4 A requirement for a minimum period of light ensures that the seeds do not germinate until the day length increases in spring;
5 Seeds that must pass through a gut ensure dispersal and a good nutrient supply for the seedlings;

Q4.12 **a** Wind; **b** Wind; **c** Wind; **d** Self-dispersal; **e** Animals; **f** Animals; **g** Water;

Q4.13 Poppy – capsule fruit 'pepperpot' on a long flexible stem is shaken by the wind; seeds are tiny to aid wind dispersal; Cherry – fleshy fruits are eaten by birds and other animals; the hard outer seed coat protects the seed if it passes through the animal gut; once deposited in faeces the seed can germinate benefiting from the fertiliser surrounding the seed; Goosegrass – hooked fruit becomes lodged in the fur of passing animals;

Q4.14 Softer;

Q4.15 Drought tolerance; temperature tolerance; salt tolerance; pest resistance; higher nutritional value; faster growth; new colours; shorter and stronger stems;

Q4.16

Coding region of PG   ATGGTTATCCAAAGG

RNA produced   UACCAAUAGGUUUCC ⎫ complementary RNA
RNA produced   AUGGUUAUCCAAAGG ⎬ strands join by H bonding

Coding region of   TACCAATAGGTTTCC
inserted gene

Q4.17 It cannot compete with other plants that flourish in the warmer conditions;

Q4.18 Their ranges would move to higher altitude up the mountain; where it is cooler;

Q4.19 **a** 31 °C;
b Isoenzymes may have different temperature optimums; they could synthesise the form with the appropriate temperature optimum as temperature changes;

Q4.20 Cooler nests producing males will become less and less common; threatening the survival of the species;

Q4.21 A straight line shows a gradual increase of about 1.8 °C between the mid-seventeenth century and the present day; a curve following trends shows fluctuations between 1700 and 1900, then a gradual rise until the present day;

Q4.22 Absence of oxygen and presence of acidic conditions reduce the activity of microorganisms; fewer survive and enzyme activity in those that do may be affected by low pH;

Q4.23 The climate was wet at the time;

Q4.24 Large amount of birch; replaced by Scots pine, elm and oak; the Scots pine disappeared; lime and alder appeared; elm disappeared leaving high percentage of oak and alder and a small amount of lime up until the present day;

Q4.25 **a** If temperature extremes were becoming more pronounced with colder winters and hotter summers; this would not be shown by an average for the whole year;
b Two cold periods 18 000–13 000 and 11 000–10 000 years ago; trend towards higher temperatures in February but with very wide fluctuations;

Q4.26 The youngest ring is at the outside and as you go deeper into the tree trunk the wood gets older;

Q4.27 **a** 900 AD and 1450 AD; **b** 1150 AD;

Q4.28 Carbon dioxide 3.7×10^{-2}; methane 3.6×10^{-3};

Q4.29 A utilitarian framework seeks to maximise the amount of good in the world; if global warming is likely to lead to more extreme weather (storms and droughts); and to certain habitats being unable to support existing flora and fauna; it is likely to increase human unhappiness and species extinctions; a utilitarian framework would therefore probably argue that attempts should be made to slow down, halt or even reverse global warming;

Q4.30 **a** 308 ppm; **b** 370 ppm;

Q4.31 **a** Photosynthesis; **b** Biomass; **c** Respiration; **d** Decay; **e** Combustion;

Q4.32 Because it merely causes carbon to be released through combustion instead of through respiration (decay); the fuel source has recently absorbed the carbon dioxide which is now being released;

Index

The authors and publishers would like to thank the following for permission to use photographs:

T = top, **B** = bottom, **L** = left, **R** = right, **M** = middle

SPL = Science Photo Library

Cover: Corbis and Digital Vision

Page 2, **T** Getty Images/Photodisc; 2, **B** Corbis; 3, Getty Images; 4, x2 Anne Scott/Mark Tolley; 5, x2 Anne Scott/Peter Kempson; 8, **T** SPL/John Radcliffe Hospital; 8, **B** SPL/Biophoto Associates; 11, **T** SPL; 11, **B** SPL/CNRI; 12, SPL/Biophoto Associates; 13, SPL/Dr Yorgos Nikas; 14, SPL/Professor PM Motta, G Macchiarelli, SA Nottola; 15, SPL/CNRI; 16, SPL/Deep Light Productions; 17, SPL/Doug Plummer; 22, John Birdsall Photography; 26, SPL/BSIP, Laurent H American; 27, SPL/Saturn Stills; 30, Harcourt Education Ltd/Peter Gould; 34, SPL/Rosenfeld Images Ltd; 35, Harcourt Education Ltd/Trevor Clifford; 36, Harcourt Education Ltd/Gareth Boden; 39, Royal Geographical Society/Ranulph Fiennes; 42, SPL/Mark Thomas; 46, **T** SPL/Sheila Terry; 46, **B** Alamy; 48, SPL/Antonia Reeve; 50, **T** Getty Images/Photodisc; 50, **B** SPL/Ian Hooton; 52, SPL/Jurgen Berger, Max-Planck Institute; 53, SPL/P Motta, Dept. of Anatomy, University "La Sapienza", Rome; 55, SPL/Biophoto Associates; 60, Dennis Kunkel Microscopy, Inc; 70, SPL/Kenneth Eward/Biografx; 74, SPL/Photo Researchers; 80, SPL/Omikron; 82, SPL/Eye of Science; 87, SPL/Hattie Young; 91, **T** SPL/Louise Lockley/CSIRO; 91, **B** SPL/Jerry Mason; 92, SPL/Saturn Stills; 93, SPL/Pascal Goetgheluck; 96, **T** Corbis; 96, **BL** SPL/Pascal Goetgheluck; 96, **BR** Nature Picture Library; 98, **T** Getty Images; 98, **B** SPL/Dr Kari Lounatmaa; 99, Biophoto Associates; 101, SPL/Don Fawcett; 102, SPL/D Phillips; 107, Getty Images/Photodisc; 108, SPL/CNRI; 109, **T**, **BL** SPL/CNRI; 109, **BR** SPL/Dr Gopal Murti; 110, x3 SPL/CNRI; 112, **T** SPL/Dr Jeremy Burgess; 112, **B** SPL/Manfred Kage; 113, **T** Nature Picture Library/Georgette Douwma; 113, **B** SPL/Science Pictures Ltd; 114 SPL/Andy Walker; 115 SPL/James King-Holmes; 116 SPL/James King-Holmes; 118, **T** Wellcome Trust Medical Photo Library; 118, **BL** SPL/CNRI; 118, **BM**, **BR** SPL/Andrew Syred; 124, Anne Scott; 126, x2 SPL/Eye of Science; 128 SPL/BSIP, Laurent H American; 130, **T** National Himalayan Rabbit Club; 130, **BL** SPL/John Eastcott & Yva Momatiuk; 130, **BR** SPL/William Ervin; 133, SPL/James Stevenson; 134, SPL/Sam Ogden; 136, **L** SPL/David Parker; 136, **R** SPL/Philippe Plailly; 139, SPL/Zephyr; 141, SPL/Mauro Fermariello; 144, **T** Getty Images/Photodisc; 144, **B** SPL/John Mead; 145, Oxford Scientific Films; 146, SPL/Frederick Ayer; 147, SPL/Alex Bartel; 148, SPL/Dr Jeremy Burgess; 150, SPL/Sheila Terry; 151, CEC/Mark N Boulton; 152, SPL/Dr Jeremy Burgess; 154, **T** SPL/Andrew Syred; 154, **B** Biophoto Associates; 155, **T** SPL/R Maisonneuve, Publiphoto Diffusion; 155, **B** Angela Hall; 158, SPL/Claude Nuridsany & Marie Perennou; 160, SPL/Biocomposites Centre/Eurelios; 161, SPL/Carlos Munoz-Yague/Eurolios; 162, SPL/Volker Steger; 163, Holt Studios; 164, **T** SPL/Dr Jeremy Burgess; 164, **B** SPL; 166, SPL/Dr Jeremy Burgess; 167, Biophoto Associates; 168, **TL** SPL/Claude Nurisany & Marie Perennou; 168, **BL** SPL/Eye of Science; 168, **R** SPL/BSIP Chassenet; 170, Angela Hall; 172, SPL/University of Cambridge Collection of Aerial Photographs; 174, **T** Department of Biochemistry & Genetics, The Connecticut Agricultural Experiment Station; 174, **B** Corbis/Ted Spiegel; 176, SPL/Sinclair Stammers; 177, SPL/Volker Steger; 178, Holt Studios; 183, Alamy; 184, NHPA/Stephen Dalton; 185, SPL/Microfield Scientific Ltd; 191, **T** GR Roberts; 191, **B** SPL/Andrew Syred; 193, **L** SPL/Richard Kirby, David Spears Ltd; 193, **R** SPL/Sidney Moulds; 201, Corbis, 208, SPL/Dr Morley Read; 210, SPL/Peter Ryan; 216, Biophoto Associates.